Reading the World and Inventing Science
The Textualized Nature of Theories

Percival S. Gabriel

Reading the World and Inventing Science:
The Textualized Nature of Theories.

Philippine Copyright © 2019 by Percival S. Gabriel. All Rights reserved. Printed in the United States of America by Createspace.com. No part of this book may be used or reproduced in any manner whatsoever without written permission from the author except in the case of brief quotations embodied in articles or reviews.

August 2019

Philosophy of Science

Cover Images

Book Cover of Galileo's Dialogue
https://la.wikipedia.org/wiki/Fasciculus:Galileos_Dialogue_Title_Page.png

Galileo's Telescope
https://www.flickr.com/photos/internetarchivebookimages/14764255884

Engraving of Sir Isaac Newton's Portrait
https://commons.wikimedia.org/wiki/File:Sir_Isaac_Newton._Line_engraving,_1748,_after_Sir_G._Kneller_Wellcome_V0004250.jpg

Kepler's model of solar system
https://commons.wikimedia.org/wiki/File:Kepler-solar-system-1.png

Gallileo demonstrating Astronomical Theories in University of Padua
https://commons.wikimedia.org/wiki/File:F%C3%A9lix_Parra_-_Galileo_Demonstrating_the_New_Astronomical_Theories_at_the_University_of_Padua_-_Google_Art_Project.jpg

Cover of Newton's Book, Optics
https://fr.m.wikipedia.org/wiki/Fichier:Opticks_by_Sir_Isaac_Newton.png

Photograph of Albert Einstein 1921 from Schumatzer
https://en.wikipedia.org/wiki/File:Einstein_1921_by_F_Schmutzer_-_restoration.jpg

News headline from *the New York Times* May 4, 1935
https://en.wikipedia.org/wiki/Albert_Einstein#/media/File:NYT_May_4,_1935.jpg

Dedicated

To my wife Lois and to my two beautiful daughters Percciloise and Philothea.

For my Mom and Dad and for my late Impo and Lola.

In memory of my writing mentor... Dr. Delfin Magpantay

Table of Contents

	Preface	i
Chapter 1	A Way to Begin…	1
Chapter 2	The World in the Eyes of Theory	9
	What to Know: The Aristotelian Lead	10
	The Beauty and the Math: from Ptolemy to Archimedes	18
	Ockham and the Medieval Interlude	26
	The Copernican Insurrection	29
	Kepler's Agreement	32
	Galileo's Success	35
	How to Know: Bacon's Methodical Disagreement	39
	Descartes's Disenchantment	42
	Validating What and How to Know	44
	Theory and Science Defined	50
Chapter 3	A Milieu too Many	55
	The Three Milieus	58
Chapter 4	The Milieu-already-constructed	69
	Textualizing Nature's Mechanics	69
	Textualizing Biological Life	88
	Textualizing Economic Relations	104
	Postscript 1	129
Chapter 5	The Milieu-under-construction	141
	Textualizing Quantum Phenomenon	141
	Textualizing Lilfe's Variations	151
	Humans as Builders of Their Own Milieu	165
	Textualizing Human Interaction in a Cramp Space	170
	Humans as Self-inventing Agents	183
	Textualizing Economic and Political Relations	188

	Textualizing the Market	196
	Textualizing Communicative Action	208
	Postscript 2	214
Chapter 6	The Milieu-of-the-text	221
	Quantum Disciples	223
	Science is Disturbance	228
	The Challenge of Quantum Philosophy on the Practice of Science	234
	The Rise and Fall of Enlightenment	260
	Interlude with Modernity	262
	The Birth and Death of Logical Positivism	270
	Popper's Logic	276
	Kuhn's Revolt	280
	Kuhn vs. Popper	284
	Lakatos's Counter-attack	288
	Popper vs. Kuhn vs. Lakatos	295
	The Post-modern Experience	300
	The Milieu-of-the-text	315
	The Tyranny of the Text	316
	Elegant Convincing	319
	The Textualized Nature of Theories	321
	Postscript 3	327
Chapter 7	...A Way to Begin Again	331
	References	335
	Index	347

Figures

Figure 2.1	Aristotle's Geo-centric Universe	15
Figure 2.2	Ptolemy's Epicycle	25
Figure 2.3	Kepler's Planetary Motion	34
Figure 2.4	Galileo's Manuscript Containing Drawings of his Observation	36
Figure 4.1	Bus Moving from D_{bus} to D'_{bus}	74
Figure 4.2	The Distance Traveled by the Bus Plotted along X and Y axes	75
Figure 4.3	The Mosquito and the Bus at Two Inertial Frames	76
Figure 4.4	A Flatlander on the Surface of a Cube	78
Figure 4.5	A Three-Dimensional Plot	79
Figure 4.6	Line that Travels along Three Dimensions	95
Figure 4.7	Lorentz Intertial Frames and Transformation	86
Figure 4.8	Characteristics of the Fly in Morgan's Experiment	93
Figure 4.9	Representation of Two Chromosomes	95
Figure 4.10	DNA Spindled in a Synapsed Chromosome	110
Figure 4.11	Marshall's Classic Supply and Demand Curve	112
Figure 4.12	Indifference Curve with the Combination of Two Products	116
Figure 4.13	Combination of Products X and Y Plotted against the Indifference Curve	117
Figure 4.14	Graph of a firm's Operation in Perfect Competition	119
Figure 4.15	Regions Favorable to Either Foreign or Local Currency	127

Figure 5.1	World-line of a Person in Space-Time	144
Figure 5.2	First Filial Generation of Crossed Peas	152
Figure 5.3	First and Second Filial Generations of Crossed Peas	153
Figure 5.4	Figure 5.4 First and Second Generations of Double Hybrids	155
Figure 5.5	5 First and Second Generations of Double Hybrid Peas	156
Figure 5.6	Crossover during Meiosis and the Resulting Daughter Cells	164
Figure 5.7	The EDSA-Quezon Ave. Intersection	171
Figure 5.8	Historical-Dialectic-Materialism in Line Plot	192
Figure 5.9	Pay-off Structure and Strategy for Peter and John	198
Figure 5.10	Zero-Sum Matrix for Tess and Mary	199
Figure 5.11	Prisoner's Dilemma Pay-off Structure	200
Figure 5.12	Application of Prisoner's Dilemma for Salesmen B and A	201
Figure 5.13	Theory of Communicative Action in a Diagram	212
Figure 6.1	The Bridge between Theory and the World	244
Figure 6.2	Flat Circle with 90° Component	253
Figure 6.3	Sphere with >90° Component	253
Figure 6.4	Trumpet Figure with <90° Component	254
Figure 6.5	An Event in Euclidian Representation	254
Figure 6.6	Three-Dimensional Coordinate System	255
Figure 6.7	Three-Dimensional Space Plus Time	256
Figure 6.8	Warped Space-Time	259
Figure 7.1	Group Picture of the 5th Solvay Conference Attendees, Brussels, Belgium, 1927	334

Tables

Table 4.1	Summary of Theories in their Deconstructive, Reconstructive and Foreconstructive Mode in Milieu-already-constructed	130
Table 5.1	Summary of Theories in their Deconstructive, Reconstructive and Foreconstructive Mode in Milieu-under-construction	217
Table 6.1	Comparing Kuhn and Popper	285
Table 6.2	Comparison between Newton and Einstein	286
Table 6.3	Comparing Popper, Kuhn and Lakatos	297
Table 6.4	Postmodernity's Offensives on Science	312
Table 6.5	Textualized Nature of Theories' Middle Ground	326

Preface

The beauty of the world that we live in comes with its understanding and the understanding of the world comes with its reading. One of the ways to read the world is through theories but understanding theories is not an easy task. I used to come home dead tired, late at night, after going through my theory class in the graduate school. I felt the hours were grueling, grinding and churning the assumptions of theories, as if we were munching nails and chewing screws. What made the load even worse was that the presenters of each theory in class found it difficult to report the subject matter. Instead of presenting the theory, they ended up talking about the author of the theory and the history of formulating it. They lost the substance of discussing the concepts and the limits that theories specifically explain. They missed out uncovering the logic and dynamics of how the statements in the theory work. In the end, they lapse into boredom, giving us more reasons to have a sleepy trip back home.

I found out that this has been a dilemma among students or even teachers of theory. It has always been difficult for them to chop theories into bits and gobble the meat. Understandably, theories are hard to chew. They are tough meat. They are a difficult stuff considering that the chefs who cooked them also had problems making the recipe easy to understand. This is compounded by the attitude that practice is more important than theory. Fieldwork is more significant than sitting on a chair and studying phenomena in the confines of a room. But what accounts for practice is theory and what has structured our society is theory. And theory is exciting to digest. But investigating and writing about theories are no simple tasks for they border on the Philosophy of Science which the postmodern attack on science is a serious subject matter to address. Overall, these concoctions of issues shaped this study and finally this book.

This project trod on a long circuitous road in two separate cities of the world (Manila and Dubai) within a long and winding timeline (2001-2016). I started writing this book in Manila way back 2001 when I was still about to court my wife Lois. It was a painstaking effort while I was teaching at the Philippine Women's University (Quezon City Campus). Yet this interest first germinated out of my desire to study Newton and Einstein, where the first few essays I wrote then were salvaged into this project. I also included the paper I re-wrote about Traffic which I submitted to Dr. Nanette Dungo (University of the Philippines) in October 2000. My first few inspirations to this project emanated from theoretical discussions with some of my mentors: Dr. Clarita Carlos, Dr. Rey Ty, Dr. Walden Bello, Dr. Nanette Dungo, Dr. Immanuel Lallana, Dr. Raul Pertierra, Dr. Carol Hernandez, Dr. Felicidad Dacayanan and Dr. Cynthia Bautista (University of the Philippines).

With this, I owe my gratitude to one of my best friends, Ruel Manseguiao who lent me his books on science, which I could not buy back then. Ate Tess Asuit also lent me some of her collections that gave me some insight. I am also deeply indebted to Lilia Cabuenos, head librarian of the Philippine Women's University (QC) and her "angels" who permitted me to freely use the library facilities and lent me books without charge even if overdue.

Preface

The next phase of this book was completed in Dubai in 2004 months after I married Lois. I spent most of my time there at the British Council and at the *Al Ras* Public Library. I am very thankful for their staff who provided me free access to books. Luckily, it was in *Al Ras* Public Library where I found the two books I was dying to get hold of: Thomas Kuhn's *The Structure of Scientific Revolution* and Karl Popper's *The Logic of Scientific Discovery*. With these two books the final chapter of this project was written.

The next phase was the partial completion back in Manila in 2006 when our first child Percciloise was born. It was at this stage when the book already took its form while the small fingers of our daughter kept on meddling on the computer's keyboard while I was typing on it. At this time, the comments facilitated by Maricor Baytion, former Director of Ateneo De Manila University Press were very valuable. The last stage was completed in 2016 when our second daughter Philothea was on her first year and whose little fingers were the ones crawling on the keyboard. It was at this time when I added the first chapter and re-wrote the fifth chapter while squeezing some free time in between classes and research days as I was already working as full-time faculty-member of the Department of History and International Studies of the University of the East. Hardwork inspired by Dr. Olivia Caoili, UE Research coordinator, Dr. Justina Evangelista, CAS Dean. Dr. Linda Santiago, Manila Campus Chancellor, Dr. Zosimo Battad, Caloocan Campus Chancellor, and Dr. Ester Garcia, UE president, to do research, write and publish papers became the blood and sweat of this book.

I won't forget my sisters Vangie, Nitz, and Helen, her husband Frank and my younger brothers Dave and Earl who have been a force behind my desire to write. The encouragement of Impo and equally the support of my mother, Florsifina and father, Ernesto to complete this project are gems I will always treasure. I owe them my life. Indeed I owe the reading the first few chapters of this book to my sister Florenda whom I kept bugging to take a look at some of the pages even late at night. I also owe her the logistical support for the initial promotion of this book. Finally, I owe everything to our Creator who gave me all these opportunities even providing me with a woman who was my research "confidant" and with whom I had an affair while writing this book ... my wife, Lois who also initially proofread this book.

Photo credits:

Figure 2.4 Galileo's Manuscript Containing Drawings of his Observation
(https://commons.wikimedia.org/wiki/File:Galileo_manuscript.png)

Figure 7.1 Group Picture of the 5th Solvay Conference Attendees, Brussels, Belgium, 1927
(https://en.wikipedia.org/wiki/Albert_Einstein#/media/File:Solvay_conference_1927.jpg) Public Domain

Chapter 1
A Way to Begin...

Brussels, Belgium... 1927... twenty-nine of the leading minds in physics and chemistry gathered to discuss and debate on one of the strangest phenomenon in the world of the minute and account for the behavior of the unseen but empirically observable as to its effect. This was the 5th Solvay Conference on Quantum Mechanics, an international gathering of scientists, which was bankrolled by a Belgian chemist and industrialist Ernest Solvay. Titled, "electrons and photons," the conference, which was organized and chaired by a mathematician and physicist Hendrik Lorentz, was witness to the big guns in physics to debate on the unique behavior of electrons which appear both as particles and waves.

On one end of the debate was Neils Bohr, the defender of Heisenberg principle which postulates that no one can know the location of an electron if one would know its velocity, and if its location is determined, no one would know how fast it is going. Where it is and how fast it is would simply be probabilistic. This concept was bitterly opposed by Albert Einstein who started the concept of quantum phenomenon with the theory of photoelectric effect which assumes that waves could act like particles as evidenced by light producing electric current once bombarded on a piece of metal. But Einstein contested the probabilistic character of quantum theory since his General Theory of Relativity could ascertain both the location of bodies in space and their speed of movement on account of the warping of space-time where these bodies move. Einstein retorted his classical counter-attack "God does not play dice," in response to the probabilistic nature of particles in the quantum world but Bohr answered back with his equally notorious response "Einstein, don't tell God what to do." (*Business Insider* 2015, Beenaker 2015 Michon 2015) In the end, Bohr won the argument and in the remaining days of his life, Einstein tried to prove that Quantum Theory was incomplete.

The amazing thing about this event was that, electrons did not suddenly become a fuzz acting like wave and then particle without us knowing exactly its velocity and location at the same time when Bohr won the argument. Bohr being victorious for Quantum Mechanics against Einstein's attacks did not make the electrons behave as the theory postulates. That would be funny. No amount of debate could make nature behave the way our argument would have led us into. The structure of nature is fixed and no amount of human debate could alter it. The physical world is out there working by itself which is detached from humans who observe and theorize about it. But the debate about nature is significant just as what the Solvay Conference had importantly proved. This simply tells us how science, though devoted to the detached observer and autonomously operating on its own, is a human enterprise and the debate is a way of elegant convincing.

A painful event, however, happened in 1633. This was the second time, Galileo had to be summoned and tried under Inquisition. As early as 1616, Pope Paul V and Cardinal Bellarmine had to serve a warning to Galileo not to discuss the Copernican theory in speech or in writing, though he was assured by the Pope that the Inquisition was not condemning him but only the Copernican idea that the sun is the center of the universe. The first trial under Inquisition came in 1625

A Way to Begin...

when a complaint was lodged against him for his publication of *The Assayer* in which the complaint alleged that the atomistic theory cannot be reconciled with the Catholic's teaching of the Eucharist. The Inquisition found him of no blame. But the 1633 complaint came with his publication of the *Dialogue Concerning the Two Chief World Systems* in 1632, where he contrasted the Ptolemaic and Copernican models of the universe, Ptolemy arguing for the geocentric universe while Copernicus supporting the heliocentric system. It was the Copernican system again with which his friend Pope Urban VIII allowed him to discuss as long as he only treated it mathematically. Galileo had to endure 18 grueling days of interrogation where he was found guilty of heresy and was sentenced to prison and religious penance. Under the threat of torture, Galileo recanted his claim of a heliocentric universe. His sentence was commuted then to house arrest until his death in 1642 ("Galileo and the Inquisition" 2015).

The point is, the universe did not stop and the earth suddenly replaced the sun at the center of our planetary system, when the Inquisitors won their argument under the threat of torture as Galileo changed his view. Neither did the sun moved to the center of the universe because Galileo had a better argument. This would be funny. But no amount of trial or debate could change the way nature works. No amount of disputation could make nature behave the way our argument would have led us into. The structure of nature is fixed and no amount of human debate could alter it. But this is how science works. Humans have to be convinced. The debates satisfy the context that science is a human enterprise. Though for the Inquisitors and the Catholic Church thereafter had comforted themselves that the counter-heretical view is correct after all, later findings would have controversially satisfied our belief of Galileo's correctness and paved the way for a more elegant convincing among us.

If science is a revolution as Thomas Kuhn argues, then it is not the planets nor the particles, nor the molecules, nor the social structures, nor the constructive meanings that are in revolution, but the humans as they try to elegantly convince each other. Scientific revolution comes with revolutionists who are the elegant leaders in their fields armed with their sterling statures needed for their elegant convincing.

But at the heart of this human activity is the weapon they come to argue for or against. It is not the planets *per se* that they are arguing, it is not the particle *per se* that they are debating, it is not the structures of meaning or behavior that they are contending, but it is the way these constructs are conceived, it is the way the ideas that represent them are structured that they come to debate and be elegantly convinced. At the center of the argument is a system of words and numbers, at the heart of the debate is a piece of text, at the crux of it all is the theory that they are fighting or rebelling against.

The aim of this study is to examine and articulate on the nature of theory where all scientific activities are centered and which are products of human enterprise. This is a study where theories will comprise the data set. Thus, instead of making the theory function as a guide in this research, theories are the objects to be studied. Instead of using theory as a guidepost, theories will be the data to be studied. Theories are human inventions that make for what science is as a human enterprise. It is through these pieces of text that the human enterprise of science builds its elegant convincing. In the course of this examination, theories will be discussed as guided by the Semantic View of theories. The Semantic View "identifies theories with certain kinds of abstract theory-structures, such as configurated state spaces, standing in mapping relations to phenomena. Theory structures and phenomena are referents of linguistic theory-formulations"

A Way to Begin...

(Suppe 2000:105). This model conceives the mapping capability of theories with the way the phenomena is mirrored.

The modern world, however, has progressed with nature and our selves being translated into text. We can read how healthy or sickly a patient is by reading his medical record. We can peruse his blood pressure, blood sugar, hemoglobin count, pulse rate; his age, weight and all other pertinent reports of him. And amazingly that piece of paper is he. The automatic teller machine will not transact business with a client unless he keys in his personal identification number. An individual would hardly get a job unless he effectively translates himself into a piece of paper or a résumé. A person cannot correspond via e-mail if he will not log on his password. Before our biological selves have been digitized, we have actually been translated into words, numbers and figures. No matter how long or short, how simple or complicated, these numbers and words are individually "we." If we would like to know how intelligent a person is, we take out his scholastic rating and take a look at his academic accomplishments or we let the child take a test and determine his IQ. Sometimes, a factor in determining how efficient and dedicated an employee is to his job is by examining his daily time record. These pieces of paper or cardboard are repositories of who we are.

In the same manner, nature can be read. The composition of water that we drink can be translated into symbols. The flow of liquid can be read as to its behavior along an obstruction. Even nature that we cannot see can be read. The atom and its sub-atomic particles can be shown through symbolic representations. How all these things began, which is a product of extrapolating physical evidence, as pushed to the limits of interpretation, is a textual presentation of what nature is and what our existence has been.

But the text, which this present project would deal with, is not simply a description of nature or human beings. It is made up of stronger stuff that has the capability to order our lives. I am talking of theories. The present project, therefore, will examine the nature of theories and advance the concept of textualization and further propose the *Textualized Nature of Theories*. But looking at the nature of the text would also necessitate the investigation of the nature of its subject. Since the subject of theories is the world, the nature of the world would then be investigated. Amazingly, however, as theories are the product of human invention, then theory by itself is also capable of structuring the world of humans. This study will not just be theoretically guided by the Semantic View of theories but will use theories as the subject of study.

Along this line, Chapter 1 lays down the development of science from Aristotle to the Age of Enlightenment. Chapter 2 plots the overall framework of the project, introducing the textual nature of theory that deals with two kinds of milieus: the milieu- already-constructed and the milieu-under-construction. Chapters 3 and 4, on the other hand, deal separately with the illustration of how these two milieus are textualized and, in the process, uncover their properties. Lastly, Chapter 5 presents the competing views of theories and science and advance the nature of theories in the milieu-of-the-text.

But though this project claims science as a textual encapsulation of nature and social relations, this work, however, presents the middle ground between science's defense and the postmodernists' attack on science. In fact, this project argues that while science deconstructs, it also reconstructs what has been disassembled and foreconstructs what could occur in the same class of phenomena. As the postmodernists argue that the grand narrative is thrown away, theories are actually narratives that do not just articulate on the phenomenon but tries to find the general fabric of these occurrences in a grand if not grander design.

A Way to Begin...

In order to argue on the middle ground between the postmodernity's assault on science and science's counter-attack, the present project proposes the *Textualized Nature of Theories* with the following assumptions:

- *The nature of the natural and social worlds suggests that they are both imbued with the properties of the milieu-already-constructed and the milieu-under-construction.*

- *The phenomena arising from either milieu are multidimensional while the theories that are hoped to account for them are theme and problem specific.*

- *Theories are text-independent entities that reflect and interpret two kinds of milieus: the milieu-already-constructed and the milieu-under-construction. This is the milieu-of-the-text.*

- *The deconstructive-reconstructive-foreconstructive capability is the heart of textualization enabling theories to capture and mirror the world.*

- *Humans are the center of theories; humans are the central figures of science; theories construct and reconstruct their world and they construct and reconstruct the world with them.*

- *The text has power, theories possess power; textualization is power.*

The following chapters will build on these assumptions.

Chapter 2

The World in the Eyes of Theory

"In the beginning was the word.... and the word was a text... so, in the beginning was theory."

Nothing in the scientific world has been made meaningful without theory and no theory has emerged without textualization. Theory is a text. Theory is the word. And science cannot proceed without theory.

Notoriety, however, has illustrated movies which depict thick-spectacled nuts, which movies call professors, sporting laboratory gowns and working on a network of glass tubes and flasks filled with bubbling liquid and picturing the nuts as working on an experiment and doing science. The depiction portrays a more subtle message that experiment makes science and science proceeds from experiments. It blares as if to show that without experiments science is a dead enterprise. But this is not the actual case. Experiments do not make science. Theories do. Experimentation does not make the endeavor scientific. Theory-building does. A scientist is not engaged in any loose experiment and out of such experiment scientific theory emerges. It is the reverse. It is theory that paves the pathway towards doing experiments. Theory is not like a billowing smoke that fumes out of smoldering mixtures of chance. Neither is science a product of wandering thought.

Theory originates from the Greek work *theoria* which for Aristotle meant deep contemplation. Aristotle differentiated it from *praxis* which means practice or doing. But the differentiation is both characterized with intellectual pursuits and did not place practice above contemplation. Aristotle simply divided the two pursuits from one which uses purely mental activity for the desire of knowing and the other with physical activity for the sake of acquiring know-how. The differentiation then is simply a matter of activity. A mathematician may be relegated to his desk in discovering knowledge without much physical activity but a physician does a lot of physical exertion doing surgery to dissect a person's ailment.

What to Know; the Aristotelian Lead

But the person, however, who gave importance to the word theory, was also the same person who began a method more path-breaking than simply contemplating. Aristotle was born in Stageria, Chalcadice in 384 BC some 55 kilometers from Thessalonica. Born to an aristocratic family, Aristotle, at the age of 18, went to Athens to pursue further studies from the great teacher Plato in his Academia. Story has it that at about age 38, he left the Academia for reasons of disenchantment after Plato's nephew Speusippus assumed authority to head the Academia upon Plato's death. Aristotle travelled to Asia Minor into the court of his friend and soon to be father-in-law Hermias of Atarneus where he was given the provisions and authority to sail into the near-by island of

Lesvos. With his friend Theophrastus, they embarked on a pioneering study of classifying living organisms.

The 1,632 square kilometers of land also bear an indention that formed a lagoon which teemed with a variety of living species. Here was Aristotle's natural laboratory. In here Aristotle meticulously observed and classified the organisms and thus began the taxonomical investigation of life. The study of life or biology was by no coincidence to Aristotle since his father was a physician to King Philip II of Macedonia. Aristotle later married Hermias's niece Pythias which Hermias had previously adopted as his daughter. The long stay in the island which Aristotle spent for his investigation and philosophical undertakings, led Pythias to bear him a daughter which he named after her. Persia eventually overtook Asia Minor and overthrew Hermias. This led Aristotle to flee to Macedonia in the court of King Philip II where he was commissioned to tutor the king's son Alexander, who would later be called the Great Alexander. At the death of Philip II and the succession of Alexander to the Macedonian throne, Aristotle went back to Athens and returned to his intellectual pursuits forming his own Lyceum. But with the death of the Macedonian king, came pro-Athenian but anti-Macedonian sentiments which led Aristotle to flee from Athens to Chalcis in order to escape from the same fate as his predecessor Socrates who was forced to drink poison at the fall of Athens to the hands of the invading Spartans (Durant 1961:8).

Plato, Aristotle's teacher, presented his thesis on the constitution of a well-ordered society, inasmuch as Plato's mentor Socrates also presented arguments on how the best form of human association could be created. Aristotle also thematized on the best form of human collectivities as he also looked up to the stars and theorized on the heaven. Aristotle challenged the prevailing ideas of thinkers of his time, which he called theories. Anaxagoras assumed that all homoeomerous bodies are elements that constitute everything (Hutchins 1952a: 334; Aristotle "On the Heavens" 303). Regarding the earth which he assumed to be the basic constitution of everything together with water and air, he criticized Xenophanes's theory that the bottom of the earth is infinite and that Empedocles assumed that the depth of the earth is endless as "endless the ample ember is" (Hutchins 1952a: 385; Aristotle "On the Heavens" 294).

In physics, Aristotle advanced his theory of motion.

> Nature has been defined as a principle of motion and change... Motion is the fulfillment of what exists potentially,... of what is alterable *qua* alterable alteration; of what can be increased and its opposite, what can be decreased, of what can come to be and pass away,...of what can be carried along, locomotion (Hutchins 1952b: 278; Aristotle "Physics" 201).

Anything that moves, however, does so in a place over time. With this he defined place as "the innermost motionless boundary of what contains" (Hutchins 1952b: 291; Aristotle "Physics" 212) and time is summed up as "a measure of motion and of being moved and it measures the motion by determining a motion which will measure exactly the whole motion as the cubit does the length by determining an amount which will measure the whole" (Hutchins 1952b: 300; Aristotle "Physics" 221).

> Everything that is in motion must be moved by something (Hutchins 1952b: 326; Aristotle "Physics" 242). Since everything to which motion or rest is natural... From this it evidently follows that coming to a stand must occupy a period of

time (Hutchins 1952b: 322; Aristotle "Physics" 239).

Aristotle's theory of motion serves as the foundation for his geo-centric theory of the then-known universe. He started out by establishing the principle of circular motion.

> It can be shown plainly that rotation is the primary locomotion. Every locomotion as we said before is either rotary or rectilinear or a compound of the two... Moreover, rotary locomotion is prior to rectilinear motion (Hutchins 1952a: 352; Aristotle "On the Heavens" 265).

A geo-centric universe presupposes the presence of revolving bodies, the circular motion that he established became the foundation of his theory.

> There is one heaven, then only and that it is unregenerated and eternal, and that its motion is regular... Since circular motion is not the contrary of the reverse circular motion, we must consider why there is more than one motion... there must be something at rest at the center of the revolving body; and of that body, no part can be at rest, either elsewhere or at the center. It could do so only if the body's natural movement is natural. Earth then has to exist; for it is earth which is at rest at the center. It is clear then that the earth must be at the center and immovable (Hutchins 1952a: 377-388; Aristotle "On the Heavens" 286- 297).

Aristotle theorized that the center of this known universe is composed of three basic elements: the unmoved earth with fire and water in it as enveloped by air. The sequence of revolving bodies or planets (Figure 2.1) around the earth is: Moon, Mercury, Venus, Sun, Mars, Jupiter and Saturn and beyond Saturn is a firmament of fixed stars (Wilson 1997: 30). What then is the shape of the heaven or the universe?

> The shape of the heaven is of necessity spherical. Again since the whole revolves palpably and by assumption, in a circle, and since it has been shown that outside the farthest circumference, there is neither void nor place, from these grounds it will follow that the heaven is spherical. It is plain that the universe is spherical (Hutchins 1952a: 378-379; Aristotle "On the Heavens" 287-288).

How about the movement of these bodies?

> We have next to show that the movement of the heaven is regular... If the movement is uneven, clearly there will be acceleration maximum speed and retardation since these appear in all irregular motions... If then its movement has no maximum, it can have no irregularity since irregularity is caused by retardation or acceleration (Hutchins 1952a: 379; Aristotle "On the Heavens" 288).

Since Aristotle assumed that these bodies move in a regular fashion, the next question to ask is, where do they derive their movement?

> Consequently the first thing that is in motion will derive its motion either from something that is at rest or from itself (Hutchins 1952b: 342; Aristotle "Physics" 257). The cause of this is now plain: it is because while some things are moved by an eternal unmoved movent and are therefore always in motion, other things are moved by a movent and are therefore always in motion and changing, so that they too must change (Hutchins 1952b: 346; Aristotle "Physics" 260).

Likewise, he challenged Anaximander Democritus and Anaxagoras's theory that the earth is flat (Hutchins 1952a: 386; Aristotle "On the Heavens" 295). At this point, theories as Aristotle referred to, against the objects of his criticisms, can be taken as simply ideas that present themselves for logical articulation.

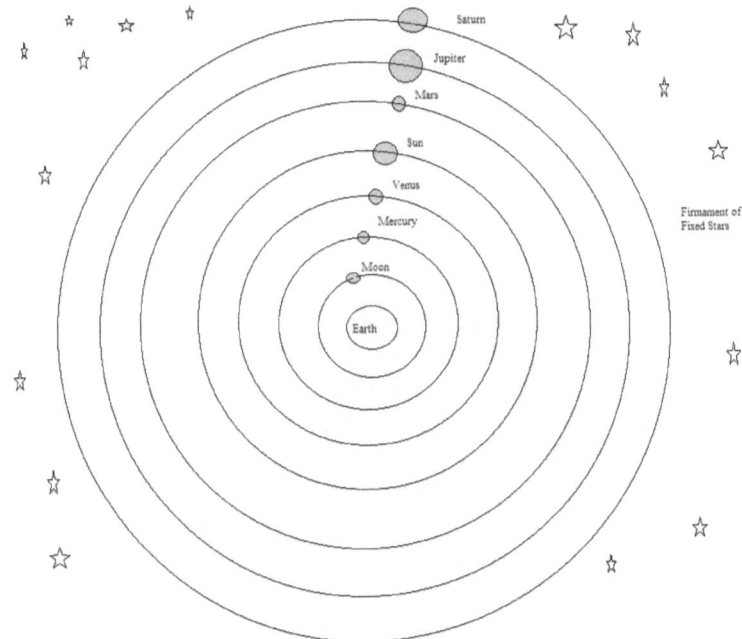

Figure 2.1 Aristotle's Geo-centric Universe

It was only of late, in 1521, when Magellan's voyage through the return of Sebastian del Cano that the earth was proven to be spherical. Aristotle's theory of the earth's shape dated back about 347 BC when he proposed the theory of a spherical earth.

> The shape of the earth must necessarily be spherical. As it is the shape which the moon itself each month shines are of every kind – straight, gibbous and concave – but in eclipses the outline is always curved; and since it is the interposition

of the earth that makes the eclipse, the form of this line will be caused by the former of the earth's surface which is there spherical (Hutchins 1952a: 388-389; Aristotle "On the Heavens" 297-298).

It was Eratosthenes of Cyrene or Libya (276 BC) who provided the measurement of the earth's circumference around 107 years later. His measurement was almost accurate as his method was anciently amazing. There were some unique observations that he found after having been appointed as the head of library in Alexandria in 240 BC. He observed that the mid-day sun strikes at the bottom of the well in Aswan. However at about the same day during the summer solstice the mid-day sun strikes at an angle for it casts a shadow on an obelisk in Alexandria. He used a plumb line to measure the angle of the shadow cast in the obelisk and found out it was 7.2°. Seven point two degrees of an angle is 1/50 of a circle's 360°, which he presumed as the earth's complete angle since he assumed it as round. Story has it that he asked some slaves to travel from Aswan to Alexandria on board an ox cart mounted with lumped wheels. Every lump on the wheel would register a certain length of measurement knowing the wheel's circumference (Holt-Jensen 1999:18). Between Aswan and Alexandria they could have probably measured around 5,040 stadia where 1 stadium is equal to 157.5 meters. Eratosthenes multiplied the distance between the two locations (5,040 stadia) with 50 being a fraction of the circumference (1/50) since Eratosthenes agreed with Aristotle's theory of regularly spherical earth. The measurement yielded the circumference of the earth to be 252,000 stadia. This is equivalent to 39,690,000 meters or 39,690 kilometers almost close to 40,000 kilometers as the circumference of the earth is known today.

These theories would lead us to how Aristotle used the word. Theories within Aristotle's usage are systematic formulations of propositional causes. Aristotle's metaphysics or the philosophical endeavor to ascertain the essence of things was geared towards knowing the four causes that constitutes the nature of a thing. These are material cause, formal cause, efficient cause and final cause. Material cause is the "stuff" that makes a thing, formal cause is its essence, efficient cause is what brought it about and the final cause is its purpose (Miller 1992:92). Aristotle's metaphysics deviates from Plato, his mentor's theory of form that assumes that the nature of things is isolated from the thing. Therefore the "idea" about a thing or its essence is separate from the thing itself. The thing can be destroyed but not the idea that brought it about. Aristotle assumes that the essence of a thing is inherent in the object. And since his theories capture the essence of these things, purporting to uncover the causes that brought it about, then theories are these formulations made up into propositional statements that are meant to uncover these causes.

He also admitted that there are various sciences. Nature, according to Aristotle's definition, is a principle of motion and change (Hutchins 1952c: 201; Aristotle "Physics" 278)... and the science of nature is concerned with "spatial magnitudes and motion and time and of each of these at least is necessarily infinite and finite" (Hutchins 1952c: 280; Aristotle "Physics" 203). Science or *dionia* is either practical (ethics and politics), poetical (study of poetry) or theoretical (physics, mathematics, metaphysics, theology) (Hutchins 1952b: 548; Aristotle "Metaphysics" 1025).

Aristotle's work was immense. The same person which began a type of scientific method by modern reckoning – naturalistic observation and rule-governed classification, he had also paved the path to formulate the means for which these intellectual propositions could be presented and that is by means of syllogistic representation which will serve as the pattern for epistemic validation in the philosophy of science that was developed in the modern times. While Aristotle provided a method for the acquisition of universal patterns among biological species in their classification, it was also Aristotle who founded the way to express them. Any expression, geared for scientific explanation, formulated outside of these syllogistic patterns becomes untrue. "Truth therefore has a structure." Scientific constructs, claims or conclusions, if they have to be deemed true have to conform to a certain structure. What is claimed to be scientifically true, should be achieved by a scientifically agreed method, must have generated scientifically proven results and should be expressed in a structurally mandated expression. This would mean that statements designed to advance scientific claims or be regarded as scientific explanation should not be true essentially or substantially but should also agree with rule-governed syllogistic presentation or form.

Such syllogistic expression should start from a universal premise, intermediated by an excluded middle statement and from which the explanation of a particular situation will be derived. Science therefore is deductive in nature. Universal claims have to be initiated first before an explanation of the particular can be derived. Thus is the need for theory as a universal claim. But how would we know that the theory is correct for an erroneous claim, though with the right form, will result in a faulty conclusion since a theory is the result of deep contemplation and not of experimentation?

The Beauty and the Math; from Ptolemy to Archimedes

With the theory of a geo-centric "heavens" or known-universe, Claudius Ptolemy (85-165 AD) fleshed in Aristotle's theory with more mathematical meat. In his book *The Mathematical Compilation*, later re-titled The *Greatest Contemplation* which in Arabic was re-named *"Al-Majisti"* and Latinized into its title *Almagest*, Ptolemy advanced his mathematical theory of the motions of the sun, moon and the planets (O'Connor and Robertson 1999).

Mathematics, however, is an area of study which Aristotle widely spoke about but did not contribute much. Mathematics was developing in a different fashion in a different direction in the hands of a different man. The same Greek as the great philosophers, Pythagoras (560 – 480 BC) whose theory of the heaven having "left a right length, right of breadth and front of depth" (Hutchins 1952a: 376; Aristotle "On the Heavens" 285) tells of his view that the heaven has three dimensions inasmuch as everything else is three-dimensional. This is understandable for Pythagoras, as all other disciples of his, believed that "all is number" (Merzbach and Boyer 2011:45) and that the essence of all things is number or that everything can be reduced into numbers. This belief is the foundation of their discovery that a ratio of three sticks 3:4:5 of whatever length forms a right triangle that will satisfy the Pythagorean formula $a^2+b^2=c^2$ (Von Fritz 1975: 220). Since all things can be reduced into numbers then musical harmony can also be done in the same way as well. They discovered that beautiful musical harmonies are created out of certain ratios of the succession of notes such as octaves (2:1) fifth harmony (3:2) and fourth harmony (4:5) (Von Fritz 1975:220). These harmonies produce beautiful sounds out of the intervals of notes. Since beauty can be reduced into number then the harmony of planetary motion can also be defined with beauty and that such beauty can be translated into numbers.

Well, Pythagoreans should have waited for centuries more in order to find out that sound is really number for sound is frequency corresponding to the number of fluctuations producing the kind of note. On the other hand, what Pythagoreans discovered were the harmonic ratios that produce the chord progression which all musical harmonies subscribe to.

The Pythagoreans developed an astronomical theory according to harmonic beauty and placed the sequence of the planets from the earth as: moon, Mercury, Venus, Sun, Mars, Jupiter, Saturn and refined it into: moon, Sun, Mercury, Venus, Mars, Jupiter, Saturn (Von Fritz 1975:223). How do these planets move also satisfy their idea of beauty and numbers.

> Pythagorean ideas of beauty required that the stars move in the simplest curves. This principle thus demanded that all celestial bodies move in circles, the circle being the most beautiful curve (Von Fritz 1975:224).

The segue to this planetary movement was proposed by Eudoxus of Cnidus (400 – 347 BC), Aristotle's contemporary perhaps, but little reference was made of him. Eudoxus wrote his book *On Speeds* which was designed to explain the irregularities of the movements of the planets as observed from the earth since planets vary in brightness as seen from the earth, implying their varied distances. Eudoxian system provided close estimates of the synodic periods of Mercury, Mars, Venus, Jupiter and Saturn (Huxley 1975:446-467).

Planetary motion is geometrical and geometry would develop in the hands of another Greek. Euclid (295 BC) wrote of his monumental book *Elements* which included proofs and definitions of figures. Mathematics, however, is unique unlike biology which Aristotle helped develop. The taxonomical classification and nomenclature of animals are readily observable but numbers and figures though readily observable cannot just be named or defined without proof. For example, a circle is observable but how can it be defined and its definition proved? Euclid agreed with Aristotle that the definition of a term must be proved or assumed.

> I call the basic truths of every genus those elements in it the existence of which cannot be proved. As regards both those primary truths and the attributes dependent on them the meaning of the name is assumed. The fact of their existence as regards the primary truths must be assumed; but it has to be proved of the remainder, the attributes. This we assume the meaning alike of unity, straight, bent, triangular, but which as regards unity and magnitude we assume also the fact of their existence, in the case of the remainder, proof is required (Hutchins 1952b: 104; Aristotle "Posterior Analytics" 766).

Euclid defined a point, a straight line and a circle as basic elements in order define and prove other geometrical figures. For example, only one straight line can be drawn from two points and the line can be extended indefinitely in both directions. But the line that bends would meet at a certain distance and will produce a circle but such circle may have infinite radius for the radius of the circle may be infinitesimally small. However, two parallel straight lines will not meet but two non-parallel lines will meet at a lesser angle intersected by another straight line (Bulmer-Thomas 1975:416-417).

Highly relying on *reduction ad absurdum* as Euclid also did, Archimedes (287-212 BC), proceeded on finding the area of figures as Euclid first defined them. *Reductio ad absurdum* is a logical means of proving a claim by proposing what it is not or its contrary and proving that it is false, in so doing, proving the claim. In his proposition:

> The area of any circle is equal to a right-angled triangle in which one of the sides about the right angle is equal to the radius and the other to the circumference of the circle" (Clagett 1975:216).

This proposition, which is today called theorem, is followed by two more statements that are contrary to the proposition.

> Let ABCD be the given circle, K the triangle described.
>
> Then, if the circle is not equal to K, it must be either greater or less" (Clagett 1975:216).

Archimedes was able to draw proof that this is true by bisecting a circle into its radius and drawing a square inside it. Now, if the two following statements are proven false, then the theorem or proposition is true. This method is particularly unique for mathematics since numbers and figures are observably existent. A triangle is never a triangle if it does not have three sides, a square is never a square if it does not have four equal sides, inasmuch as a circle is never a circle if its radius is not equal in all directions. These statements concerning these figures are true because there are no other statements that are true to the contrary. They are intrinsically true. These are axioms. But measuring their component angles, finding their properties, ascertaining their areas, cannot simply be reduced into statements by simple observation like Aristotle's taxonomical classification through observation. They have to be proven by other statements through acceptable methods. These statements or propositions that are proven are theorems.

Now, established within this mathematical tradition, is Ptolemy's mathematical theory which provided the mathematics for Aristotle's geo-centric universe. For Ptolemy, nothing can be farther from the truth as nothing can be far from what is observable, he being a mathematician. Stand and look up the sky during the day and at night and you would observe that everything in the heaven revolves around you as the sun rises in the east and sets in the west, the moon and the stars also seem to move on the dome of the heaven. At this time, one whole day had a crude way of measurement. One whole day can be divided into light (daytime) and darkness (night time). That is, you can witness the sun rise and set at daytime and see the moon shine at night time. Since no apparent astronomical observation can be done during daytime, the moon provides a clue on the length of the month. The cycle of the phases of the moon from new moon to full moon would yield exactly 30 days or 30 sunrises or sunsets. This is equal to a month. Seasons, however, change and this is subject to the changes of the locations of the stars in the night sky. Count the number of months in the cycle of the seasons and it will give off 12 months. And that completes a year. All these calculations are dependent upon the locations of the heavenly bodies in the night sky or at day time. No wonder, these philosophers and mathematicians are fascinated with the heaven since the conditions on earth are dependent upon the motions of these bodies in the sky. But there was something still left without measurement though the months and years have already been divided. The day can be arbitrarily divided into any number of separations. The Egyptians, however, were the first ones to divide the day and night into 24 subdivisions, 12 for the day and 12 for the night (Wilson 1997:19). One of the twenty-fourth (1/24) of the day is called an hour.

These measurements served as the basis for Ptolemy's mathematical calculations of an earth-centric universe. But there are anomalies which he needed to account for. The length of the day and night changes through the seasons though we divide the day and night into 24 separations. Since Ptolemy agreed with Aristotle's geo-centric universe and the movement of the planets are regular around the earth and such movement is a circle since it is a model of beauty, the longer days can be explained by longer lengths of revolution as the sun rotates around the earth at certain times of the year. Longer lengths would mean longer radius of revolutions around the earth. It was only during the solstice months that equal lengths of the day and night were observed.

> In addition to those already mentioned, this general assumption would also be rightly made that there are two different prime movements in the heavens. One is that by which everything moves from east to west, always in the same way and at the same speed with revolutions in circles parallel to each other and clearly described about the poles of the regularly revolving sphere... The other movement is that according to which the spheres of the stars make certain local motions in the direction opposite to that of the movement just described and around other poles than those of that first revolution... The one first movement which contains all the others will be thought of then as described as if defined by the great circle, through both sets of poles which is carried around and carries with it all the rest from east to west about the poles of the equator... But the second movement, consisting of many parts and contained by the first, and embracing itself all the planetary spheres is carried by the first as we said, and revolves about the poles of the ecliptic in the opposite direction (Adler 1993c: 13-14).

In order to resolve the anomaly, Ptolemy still assumed that the sun revolved at an equal radius around the earth, making the movement regular, but the sun wobbled in small revolutions as it did. These small revolutions in a revolution is like stretching a spring and these small wobbles in the entire spring that can be coiled in a circle are, as Ptolemy called, epicycles (Figure 2.2).

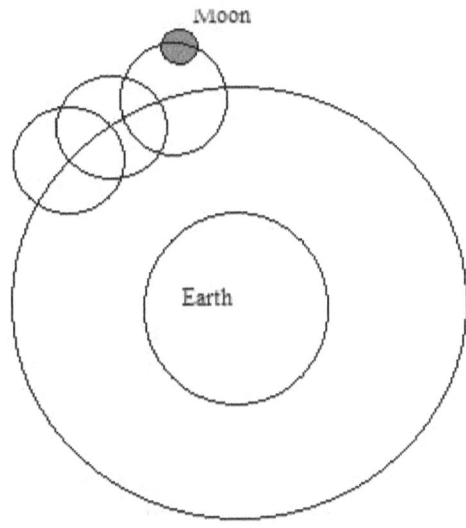

Figure 2.2 Ptolemy's Epicycle

Earlier astronomers... recognized that planetary motions could not be represented, ever crudely by a simple circular motion about the earth and had proposed a combination of two circles, one in a large circle (deferent) and one in a smaller circle (epicycle). The center of the epicycle moved around the epicycle at a constant velocity and the planet moved around the epicycle at constant but different velocity (Wilson 1997: 35-37).

Ptolemy adopted this technique and developed the trigonometry that he used which was called a chord function, related to the sine function ($\sin \alpha = [\text{Crd } 2\alpha]/120$) (Connor and Robertson 1999 and Toomer 1975:188-190).

The relative success of Ptolemy's mathematical model to give mathematical substance to Aristotle's geo-centric theory is a testament that mathematics can be pliantly manipulated in order to correspond and explain a phenomenon. This means that whether the earth is the center of the system or not, mathematics can be made to fit into the observables. Mathematics then is a descriptive tool. This goes against the true Pythagoreans who subscribe to the idea that everything is number or that number is essentially the nature of things. Consequently, this produced a way of looking at things. Contrary to the Pythagoreans that everything can be reduced into number, mathematics and numbers are only a way of looking at the "appearances" of things or what can be observed. This is called "saving appearances" (Losee 2001: 17-18).

Aside from the philosophical view that numbers are simply meant to describe what appears, is the careful interpretation of the heaven for thinkers were perceived to be treading on dangerous grounds. Aristotle admitted that "the ancients gave to God the heaven or upper place as being alone immortal" (Hutchins 1952a: 375; Aristotle "On the Heavens" 285). Aristotle was even careful of his theoretical view of the heaven. His known universe was only limited to Saturn and the firmament of fixed stars for beyond the firmament could be the dwelling of the gods and he could be treading on the thin line of blasphemy.

Ockham and the Medieval Interlude

While reason made its headway to discover the world during the Greeks, reason, however, took a back seat during the medieval age. The Catholic Church that emerged as the empire's religion during the Roman period survived the barbaric carnage that sacked and ruined Rome. It continued to become the power broker among monarchs that ruled kingdoms in Europe. Its power to broker with kings came with the Vatican's capability to legitimize the rule of monarchs by investing upon them the public recognition of the Divine. But other than power brokering for political ends, the Catholic Church also dictated on what should be accepted as episteme.

Knowledge came at the bridle of the church but at the expense of reason. The Greek classics such as Aristotle's writings were lost in the carcasses of knowledge that the church mangled. But the institution that assaulted it also helped in its rescue. The monasteries discovered Aristotle's texts and influenced late medieval writers such as St. Thomas Aquinas who accepted the Aristotelian thesis of happiness as the end of society (Curtis 1981: 177). Reason re-emerged but this came with a bargain. Reason was accepted as human's essential nature but its product had to conform with the teachings of the Catholic Church. Aquinas, in fact, fused human reason with the divine in his *Summa Theologica*, arguing that "law as the dictate of practical reason" summed up human legislation that should find connection with the natural law and the divine law (Curtis 1981:200-205). This would mean that human law as an expression of human will and reason should have direct connection with the divine will.

Another philosopher during the 13th century was committed at the marriage of reason and faith. William of Ockham argued that God can produce anything without self-contradiction for anything that is self-contradictory deprives itself of truth. This knowledge comes to us through experience.

> Except for premises of mathematics, which are known *per se* by the meanings of the terms, the principles of the natural sciences are held by Ockham to be evident by experience but not necessarily in the absolute sense, although they may be said to be necessary in the conditional sense of presupposing the common course of nature without Divine interference (Moody1974:174).

Thus with the first maxim he says: "Nothing is to be assumed evident unless it is known *per se*, or is evident by experience, or is proved by the authority of the scriptures" (Moody 1974: 173). And in terms of scientific knowledge, his second commitment is expressed in his principle known as Ockham's razor that: "What can be accounted for by fewer assumptions is explained in vain by more" (Moody 1974: 173). Ockham's razor slices through different claims or competing theories where simplicity and economy of assumptions to explain the phenomenon is the right theory to explain it. Ockham used his razor to cut through the medieval debate on the projectile motion.

> One view was that a projectile's motion is caused by an acquired "impetus" which resides somehow in the projectile as long as it is in motion (Losee 2001:34).

But motion, was believed, not to be a property of a body, but produced in relation to other bodies at a certain period of time.

> Ockham maintained that to say a 'body moves because of an acquired impetus' is to say no more than 'a body moves' and he recommended the elimination from physics of the concept of "impetus" (Losee 2001:34).

The statement, therefore, that projectile motion is caused by an acquired impetus is redundant. In that case, Ockham is showing that scientific explanation follows a simple path and that economy and simplicity bring about truth. While Ptolemy and Euclid believed that nature follows beauty as its path, Ockham claimed that giving account of nature follows a simple and economic corridor.

The Copernican Insurrection

With the revisiting of Aristotle in the late medieval age came the re-surfacing of Greek science. And with it was the idea of geo-centric universe which Ptolemy gave mathematical flesh. But an astronomer and mathematician found dissatisfaction with Ptolemy's calculations. Nicolaus Copernicus, a Polish, published his book *De Revolutionibus* (On the Revolutions) in 1543 where he wrote his radical departure from Aristotelian universe and Ptolemaic calculations.

> Aristotle says that the movement of a body which is one and simple is simple, and the simple movement are the rectilinear and the circular. And of rectilinear movement, one is upward, and the other is downward. As a consequence, every simple movement is either toward the center, i.e. downward, or away from the center, i.e. upward, or around the center, i. e. circular. Now it belongs to earth and water, which are considered heavy, to be borne downward, i.e. seek the center: for air and fire, which are endowed with lightness, move upward, i.e. away from the center. It seems fitting to grant rectilinear movement to these four elements and to give the heavenly bodies a circular movement around the center. So Aristotle, and therefore, said Ptolemy of Alexandria, if the earth moved, even if only by its daily rotation, the contrary of what was said above would necessarily take place. For this movement which would traverse the total circuit of the earth in twenty-four hours would necessarily be very headlong and of an unsurpassable velocity (Adler 1993b: 518).

The Aristotelian universe begins with its constitution as earth, water, air and fire. At the middle of the universe is the earth and everything moves laterally depending on its heaviness or gravity, thus air and fire moves upward because they are light. With the earth at the center, this universe consists of other revolving bodies as the moon, Mercury, Venus, Sun, Mars, Jupiter and Saturn and beyond Saturn is a regime of fixed stars. All these bodies are observable in the naked eye like stars but even Ptolemy and his contemporaries observed that these bodies did not uniformly move in a circle around the earth just as what Aristotle theorized while preserving the maxim that a circle exemplifies perfection and beauty and through which nature obeys. Even Ptolemy observed that the movement was not equal at the center. Planets would circle away a little and then continue on with the orbit. Ptolemy resolved this by introducing the concept of epicycles or a wobbling small orbits around a bigger orbit like the path of a stretched spring around the earth. The epicycles would then produce different centers or equants. But just as Ockham said, there are things unnecessary in this to account for the constitution and the movement of the universe.

> ...the ancients allowed the epicycle to move uniformly only around the equant's center. This procedure was in gross conflict with the true center [of the epicycle's motion], its relative [distances], and the prior centers of both [other circles]...However, in order that this last planet too may be rescued from the affronts and pretenses of its detractors, and that its uniform motion, no less than that of the other aforementioned planets, may be revealed in relation to the earth's motion, I shall attribute to it too, [as the circle mounted] on its eccentric, an eccentric instead of the epicycle accepted in antiquity (Rabin 2010).

In order to straighten up this contradiction and simplify things, Copernicus placed the sun at the center and the planets moving around it. The geo-centric universe of Aristotle and Ptolemy was not just challenged but was demolished.

Therefore if the first law is still safe – for no one will bring forward a better one than that the magnitude of the orbital circles should be measured by the magnitude of time – then the order of the spheres will follow this way... Saturn, the first of the wandering stars follows; it completes its circuit in 30 years. After it comes Jupiter moving in a 12-year period of revolution. Then Mars, which completes a revolution every 2 years. The place fourth in order is occupied by the annual revolution in which we said the Earth together with the orbital circle of the moon as an epicycle is comprehended. In the fifth place, Venus, which completes its revolution in 7 ½ months. The sixth and final place is occupied by Mercury, which completes its revolution in a period of 88 days. At the center of all rests the sun. For who would place a lamp of a very beautiful temple in another or better place than this wherefrom it can illuminate everything at the same time? (Adler 1993b: 526-527).

While these planets move around the sun, including the earth, the moon moves about in a different path as it orbits around the earth. Copernicus, however, may have had some inspiration from other writers since Aristarchus of Samos wrote that the earth revolved around the sun, inasmuch as Pythagoras also believed that the earth circled around a central fire (Rabin 2010). The radical theory of a helio-centric universe was a huge departure to the Church's geo-centric universe which would put Copernicus in trouble for the church espoused the Aristotelian universe of an earth-centric system. The Church, however, was without any defender. Those who believed that there was no contradiction between faith and reason viewed the Copernican revolution as simply an appearance and not an essential attribute of the universe. This would mean that Copernicus' mathematical calculations are simply attributes and not the universe's innate property.

Copernicus, however, was a Pythagorean who viewed these calculations as the universe's essential nature. In fact, to rescue such mathematical calculations as simply "appearances" and not properties, a Lutheran theologian, Andreas Osiander who wrote the preface of *De Revolutionibus* wrote that "it does not matter whether the planets really do revolve around the sun, what matters is that Copernicus was able to save the appearances of the assumption" (Losee 2001: 40).

But despite such a departure from Aristotle's universe and the Church's doctrine, there is one thing that Copernicus still preserved. He adopted the aesthetic doctrine that nature obeys a circular movement around the central fire. The planets still moved in perfect uniform circular movement around the sun.

Copernicus' *De Revolutionibus* was a rebellion in itself during the Age of Enlightenment in Europe which did not just bring about scientific revolutions but also rebellion against the Church's epistemic impositions. This revolution did not just spur debates on the natural sciences but also questioned the Church's dictates on brokering power to create a society and in turn championed reason and human will for the people to organize their society.

Kepler's Agreement

Against Copernicus' detractors and against the inconsistencies brought about by his notion of the circular motion of the planets, Johannes Kepler, a German mathematician, who was a true Pythagorean saved Copernicus from the gaps that a circular motion generated. Kepler published his most famous work *Mysterium Cosmographicum* in 1594 where he advanced that the distances of planets can be correlated with the radii of spherical shells. He worked with

Tycho Brahe from 1597 until Brahe's death in 1601. It was in 1609 when he published *Astronomia Nova* and in 1619 *De Harmonice Mundi*, two of his famous works that comprised his theory of planetary motions which was later recognized by other scientists as laws of planetary motions. Kepler assumed that:

> (1) The orbit of a planet is an ellipse with the sun at one focus.
>
> (2) The radius vector from the sun to a planet sweeps over equal areas in equal times.
>
> (3) The ratio of the squares of the periods of any two planets is directly proportional to the ratio of the cubes of their mean distances from the sun (Losee 2001: 43).

Kepler's theory with the first statement or law agrees with Copernican planetary system of a sun-centered universe but it hugely departs from circular motion of the planets where equality in terms of the distance to the center is preserved. Kepler believed that planets did not travel around the sun in perfect circle but revolved around the ball of fire in an elliptical orbit. What Kepler preserved with its equality is the area of sweep as the planets move around the sun. The elliptical path of a planet around the sun sweeps equal areas at equal times. With S at the center (Figure 2.3), the area that planet P sweeps at an arc-distance from a to b is equal to the area that it travels from the arc-distance from c to d. And travelling at c to d is faster at the foci S than negotiating a to b. This would mean that the planet would travel faster on the area closer to the foci or the perihelion and slower in area away from the foci or the aphelion of the ellipse. The third statement or law can be stated this way: the square of the planets revolution around the sun or period (T^2) is directly proportional to the cube of its average distance from the sun

(R^3). Thus with respect to two planets the proportionality would be: $T_1^2/R_1^3 = T_2^2/R_2^3$.

> Then what is that true movement of the planets through their surroundings? It is constant with respect to the whole periods; and proceeds around the sun, the centre of the world, always eastward towards the signs which follow. It never sticks in one place, as though stationary, and much less does it ever retrograde. But nevertheless it is of irregular speed in its parts; and it makes the planet in one fixed part of its circuit digress farther far from the sun, and in the opposite part come very near to the sun: and so the farther it digresses, the slower it is, and the nearer it approaches, the faster it is. Finally, in one part of the circle it departs for the ecliptic to the north, and in the other, to the south, and so the planet is left with a real irregularity and one which is threefold, too (Adler, 1993a: 928).

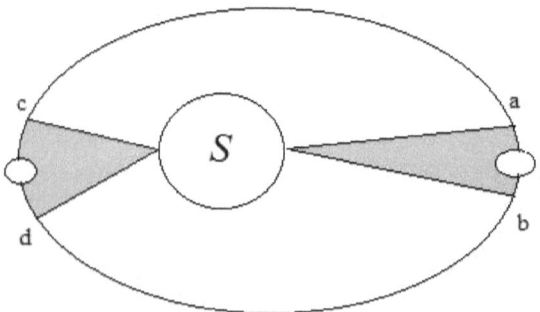

Figure 2.3 Kepler's Planetary Motion

Kepler's method was a first since he was able to use voluminous data observed and recorded by his predecessor Brahe and made sense to all of them into mathematical statements that are essentially the properties of nature just as what Pythagorean believed.

Galileo's Success

But a further breakthrough was achieved by an Italian Galileo Galile. While others still held on to earth's centricity and still believed that Kepler's theory was only hypothetical, Galileo was able to find physical evidence of the revolutionary idea. Story has it that he was mesmerized by the chandelier in the church that swung like a pendulum and would stay at rest after an external body made it move. Story also has it that he stood atop the Leaning Tower of Pisa and tried throwing objects of different weights. But as early as 1609, he had begun observing the night sky with his invented telescope. Being a firm believer of the Copernican system, he set his sight on Jupiter where he found moons dotting the huge planet which created different patterns as he observed it at various times (Carnegie Science 2018). This led him to believe that these moons orbited Jupiter (Hawking 1988:4). If smaller bodies like Jupiter's moons orbited the bigger body Jupiter, how much more would the planets, smaller in size as the sun, move around the biggest body of all which is the sun. He published this discovery in 1932 in his book *Dialogue Concerning the Two Chief Systems*, where Copernicus and Kepler found most convincing proof but which harbored the indifference of the Catholic Church that still defended the earth-centric universe. The animosity bred his persecution. For the Catholic Church, the explanation was simple: the earth is the center of the universe for the earth is the home of God's supreme creation - human beings. The church found Aristotle as its proof.

Reading the World and Inventing Science

Figure 2.4 Galileo's Manuscript Containing Drawings of his Observation

But with the myth of him standing atop the Tower of Pisa and throwing bodies of different weights, it was documented that Galileo used an inclined plane to roll off balls of different weights mimicking bodies that fall to the earth. Galileo's measurement resulted to his discovery that the speed of each ball, no matter the weight, increased at the same rate (Hawking 1988: 15-16). This would mean that bodies of two different weights would fall at the same time because of the same attractive pull acting on both of them. Since they would touch ground at the same time, then the rate of increase in speed would also be the same. He published the law of free falling bodies and rescued his Copernican belief in his succeeding book, *Discourses of the Two New Sciences* in 1638. Galileo's work did not just begin modern physics but it also innovated how nature can be known or discovered. While proving the helio-centric universe, he championed that experimentation is a criterion for discovery and that scientific knowledge can be attained through instruments fitted in the experimental design.

This breakthrough led Francis Bacon to challenge the Aristotelian science. Remember that it was Aristotle who began the methodic analysis of things that impinge on our desire to know. Aristotelian science, however, is dedicated by two things: cause and form. What makes science is the search for the cause of things and its proof should subscribe to syllogistic form.

> We think we have scientific knowledge when we know the cause, and there are four causes: (1) the definable form, (2) an antecedent which necessitates a consequent, (3) the efficient cause, (4) the final cause. Hence each of these can be the middle term of a proof... (Hutchins 1952b: 128, Aristotle "Posterior Analytics" 94).

Aristotle also demonstrated the form of proof in various phenomena:

> Thus: let C be the planets, B proximity, A not twinkling. Then B is an attribute of C, and A – not twinkling – of B. Consequently A is predicable of C, and the syllogism proves the reasoned fact, since its middle terms is the proximate cause. Another example is the interference that the moon is spherical from its manner of waxing. Thus: since that which so waxes is spherical, and since the moon so waxes, clearly the moon is spherical. Put in this form, the syllogism turns out to be proof of the fact, but if the middle and major be reversed it is proof of the reasoned fact; since the

moon is not spherical because it waxes in a certain manner because it is spherical. (Let C be the moon, B spherical, and A waxing) (Hutchins 1952b: 107-108, Aristotle "Posterior Analytics" 78-79).

Put it in another way, we could deduce what Aristotle was trying to say.

All bodies in the heaven that waxes (or concaves at different times) are spherical.

The moon waxes.

Therefore, the moon is spherical.

Previously deduced from Aristotle that with his metaphysics is the idea that "truth has a structure." Syllogism is this form of presenting truth. Outside of this structure is a false claim. Aristotle did not just discover the nature of things that he observed, he also invented the way of presenting the way they could be logically and truthfully accepted.

Aristotle's syllogistic technique leads to one consequence. You cannot proceed to a conclusion without figuring out what the universal is. You cannot move on to the explanation of a particular if you have not yet made up the universal. The method calls for composing the first statement of the syllogism, injecting the particular which begs for explanation and the conclusion proceeds from it. The method, therefore, is deductive in nature. But how do we know that the universal is correct. Aristotle believed that theory is a product of "deep contemplation" but that does not make for science that we simply admit scientific knowledge to primarily originate from deep contemplation. How do we know that the product of a deeply contemplative activity is right? What is needed is proof. Galileo's experiment, however, led to proof that a free-falling body increases speed at the same rate. Galileo's observation of the heaven led to proof of the

Copernican system. Experiment, therefore, achieved success rather than simply deep contemplation.

How to Know; Bacon's Methodical Disagreement

It was with this theme that Francis Bacon aired his objection of Aristotle's science. The British lawyer and parliamentarian published his *Novum Organum* (1608-1620) which disputed Greek science.

The sciences we possess have been principally derived from the Greeks; for the additions of the Roman, Arabic, or more modern writers are but few and of small importance. As such they are founded on the basis of Greek invention (Hutchins 1952: 117, Bacon "Novum Organum").

Bacon argued that the acceptance of the Greek science brought with it the creeping in of fallacies or distortions of human understanding into the mainstream scientific investigation. He called them Idols. Just as God ordered the idols to be demolished during the time of Moses in the Bible, so too Bacon admonished the idols of human understanding be destroyed. These idols include: the Idol of the Tribe which consists of false concepts due to "prejudices or the limited faculties of humans, incompetency of the senses or interference of passion"; the Idol of the Den which are doctrines that "owe their birth from either to some predominant pursuit, or to an excess in synthesis or the extent or narrowness of the subject"; the Idol of the Market which are false concepts that "have entwined themselves around the understanding from the association of words and names"; and the Idol of the Theater which are "theories or religious doctrines or even false philosophy" (Hutchins 1952: 111-113, Bacon "Novum Organum").

Demolishing these idols would mean starting out fresh in the pursuit of scientific discovery and that would pose a challenge on the deductive technique of Aristotelian science for Bacon argued that scientific activity is inductive and not deductive. With this, Bacon also disputed the Aristotelian tradition of seeking after causes and championed Democritus in search of the basic constitution of things rather than their causes. The inductive method would mean recasting the whole scientific activity divorced from the Aristotelian contemplative mode towards syllogistic expression into laying down empirical observations towards experimentation. Experimentation comes into the methodic debate due to Galileo's experimental success in discovering and explaining nature's uniform acceleration of free falling bodies. The inductive method would incur a different form.

> The investigation of forms proceeds thus: a nature being given, we must first present to the understanding all the known instances which agree in the same nature, although the subject matter be considerably diversified. And this collection must be made as a mere history and without premature reflections, or two great degrees of refinement (Hutchins 1952: 140, Bacon "Novum Organum").

Bacon used this method to investigate heat. First, a series of instances about heat was presented in a table.

Table I. Instances agreeing the Form of Heat

(1) The rays of the sun, particularly in summer, and at noon.

(2) The same reflected and condensed, as between mountains, or along walls, and particularly in burning mirrors.

(3) Ignited meteors.

(4) Burning lightning

(5) Eruptions of flames from the cavities of mountains

(6) Flame of every kind

(7) Ignited solids

(8) Warm or heated liquids...(Hutchins 1952: 140-141, Bacon "Novum Organum")

The next stage was experimentation.

The reflection of the solar rays in the polar regions is found to be weak and inefficient in producing heat... Let the following experiment be made. Take a lens, the reverse of a burning-glass, and place it between the hand and the solar rays, and observe whether it diminishes the heat of the sun as a burning glass increases it... Let the experiment be well tried, whether the lunar rays can be received and collected by the strongest and best burning-glasses, so as to produce even the least degree of heat. But if that degree be, perhaps, so subtle and weak, as not to be perceived or ascertained by the touch, we must have recourse to those glasses which indicate the warm or cold state of the atmosphere, and let the lunar rays fall through the burning glass on the top of this thermometer, and then notice if the water be depressed by the heat (Hutchins 1952: 142, Bacon "Novum Organum").

The next stage is the exclusion or rejection of results obtained in the experiments. This would lead to the conclusion of what heat is.

On account of the sun's rays, reject elementary (or terrestrial) nature.

On account of the ready application of heat to all substances without any destruction or remarkable alteration of them, reject destructive nature of the violent communication of any new nature.

On account of the agreement and conformity of the effects produced by cold and heat, reject both expansive and contracting motion as regards the whole (Hutchins 1952: 150, Bacon "Novum Organum").

Bacon's reliance on the inductive technique led him to admit that the objective of this scientific method is the discovery of the laws of nature. Laws that have been derived out of propositions and confirmed through experimental activity are in no way the same as axioms which are statements generally accepted as true without any experimental or methodical activity. Axioms have been anciently used in geometry, like "all circles have 360° angle" need no confirmation at all.

Descartes's Disenchantment

But as Bacon sounded off his rejection of Aristotle's method, Rene Descartes also declared his dismay over Aristotelian science on account of Aristotle's method of deduction. Aristotle's syllogistic analysis was short of real proof according to Descartes who mathematicalized geometry by combining algebra with geometry in an X-Y (Cartesian) axis. Descartes disagreement stems from his argument that the universal statement or major premise has to be known first before the proper syllogism could proceed. But what method could unlock the universal or how could one prove that the universal is true except that Aristotle does not prescribe any other method except his syllogistic technique.

Descartes advanced his technique of geometrical demonstration which he found success in finding proof of mathematical concepts in his work *Rules for the Direction of the Mind* which he began writing by 1618. Descartes argued that the triumph of the technique originates from the use of axioms which are statements generally held as true and needs no further proof while also utilizing generally accepted definitions. The technique gets rid of sensation as a means to obtain knowledge for Descartes argues that this type of knowledge can be achieved only through intuition (Skirry 2008).

> Concerning the objects presented to us we should investigate, not what others have thought nor what we ourselves conjecture, but what we can intuit clearly and evidently or deduce with certainty, since scientific knowledge is acquired by no other means (Ariew 2000: 5).

Descartes then championed intuition and deduction as a means to scientific knowledge. He admitted that science is devoted to the discovery of laws of nature, laws that stand as irrevocably and inherently true to all phenomena of the same class, but that which needs confirmation by methodical enterprise. But as a believer that innate knowledge exists, he also advanced that laws are derivable first through the formulation of general statements and out of which proceed propositions that necessitate deductive verification. Thus, as divorced from Bacon who also aspire the discovery of the laws of nature through his inductive method, Descartes believes that the summit of his deductive method would redound to his dictum, *cogio, ergo sum*, "I think, therefore I am." But as Descartes championed rationalism on the one hand, a rival school of thought regarding the attainment of knowledge was ready to challenge it on the other.

Validating What and How to Know

While Galileo's experiment proved that there is only one force attracting bodies that fall on the ground, Bacon, on the other hand, tried to institutionalize the experimental method through an inductive process. But one would wonder how one could come up with the initial statements during the induction process and how could one judge upon the experiments except that they would all pass through our senses. All the inductive statements are observations and even in experiments would use instruments, the measures and results of which are all products of sense impressions. In 1687 Isaac Newton published his *Philosophiae Naturalis Principia Mathematica* which provided more victory for the use of experimentation and mathematics in the discovery of the natural phenomenon. Against the propositions of rationalists, John Locke's monumental work *An Essay Concerning Human Understanding* published in 1689 argued that the mind is an empty chest and the mind discovers or receives knowledge through the senses. Locke attacked the concept of innate ideas that indicates 'there lurks ideas which the mind, through reason, can discover.'

> Idea is the object of thinking. All ideas come from sensation or reflection... The impressions then, that are made on our senses by outward objects, that are extrinsical to the mind, and its own operations, about the impressions, reflected on by itself, as proper objects to be contemplated by it, are, I conceive, the original of all knowledge, and the first capacity of human intellect, is, that the mind is fitted to receive the impressions made on it; either, through the sense, by outward objects; of by its own operations, when it reflects on them (Winkler 1996: 33-39).

He argued that the mind is empty and can receive simple or complex ideas. Complex ideas are of two types, substance and mode. Substances are ideas with independent existence like humans, God, animals, plants while modes are ideas that are dependent like mathematical, moral, political principle (Uzgalis 2012).

If knowledge is achieved through the senses, then duality exists between the materially sensed, and the faculty of the senses that would manufacture knowledge out of them. This would result to the dual independence of material things and the rational mind that sense them. George Berkeley, an Irish Bishop attacked this duality even attacking the idea of materialism of his time that only material things exist.

Berkeley published his *Treatise Concerning the Principles of Knowledge* in 1710 where he argued that matter exists except in the mind; that the mind is not independent of matter and that knowledge of matter comes from the mind. He couched this in his famous phrase, *"esse est percipi (aut percipere)* – to be is to be perceived (or to perceive)" (Downing 2011).

David Hume, an English philosopher, salvaged his defense of empiricism in his two famous works *A Treatise of Human Nature*, completed in 1740 and *Enquiries Concerning Human Understanding* published in 1748. Hume championed the experimental method which greatly found success in the discovery of physical laws though the works of Newton on motion and gravity and Robert Boyle on pressure and heat. He suggested the use of experiment in the creation of what he called the science of human nature (Norton and Norton 2000:114). In defense of empiricism, Hume argued that there is nothing that humans would know except through sense impressions of matter. There is nothing that may have led us to know or made us think except for things that have caused our faculties to sense or feel.

> All the perceptions of the human mind resolve themselves into two distinct kinds, which I shall

call IMPRESSIONS and IDEAS. The difference betwitxt these consists in the degrees of force and liveliness with which they strike upon the mind, and make their way into thought or consciousness. Those perceptions, which enter with most force and violence, we may name impressions; and make their first appearance in the soul. By ideas I mean the faint images of those in thinking and reasoning; such as, for instance, are all the perceptions excited touch, and excepting the immediate pleasure or uneasiness it may occasion (Hume 2000:7).

By this Hume means that impressions include all sensations or emotions that enter our mind and ideas are faint images retained by experience which the mind can mine.

The method for scientific discovery had led its way to sophistication rather than simply contemplation and syllogizing. But there is a seeming triumph to this story. When Ptolemy invented his chords to provide mathematical flesh to the epicycles of planets revolving in a geocentric system, a sort of warning was sounded that mathematics is external to nature and that the formulae are simply meant to describe the phenomenon. Mathematical formulae are not nature's essence. The scenario even got worse when Copernicus mounted a revolt to propose a helio-centric universe, the more that mathematics was believed to be a descriptive mirror and not the real means of how nature operates. But when Kepler came in defense of the Copernican system with his mathematical formula, when Galileo achieved success in the same attractive force that the earth exerts using mathematics and confirmed the helio-centric system using his telescope, when Newton worked on his inertial systems using mathematics that found irrevocable success in nature, mathematics made its revenge that it is not simply a descriptive means, but it is exactly how nature works. The mathematics that explains the operation of nature is actually its essence. But there is certain uniqueness to this. Mathematics, more specifically geometry, has been a product of rational success without observation or sense necessities to find truth in its operation. Physics, however, has progressed through empirical evidence, having found satisfaction out of observation and experimentation as the product of sense impression.

But there is a side story to this rationalist-empiricist debate. Where is God in the picture? It would be recalled that the science that we are talking about is European in residence and Europe was plunged into the carousel of divine explication to things that reason would have discovered. But the idea of the Divine survived even after the medieval age, challenging the residue of a system where the centerpiece of God's creation resides. It is decreed by the Catholic Church that the earth is the center of the universe. This idea goes very well with the Aristotelian and Ptolemaic universe. Anything that goes against it was heretical. Copernicus got his first skirmish with the idea. Kepler supported it despite the anticipated Church's objection and Galileo was subjected to Inquisition for adhering to the Copernican system despite the 1616 Papal decree declaring the system false and erroneous. Twice, however, did Galileo recant his theory only to mount his revenge by publishing his last book before his death in 1642 *Two New Sciences* where he vigorously re-affirmed his defense of the Copernican system than ever before (Hawking 1988:190).

When the empiricists advanced their attack, the position of God was all the more placed in a precarious position. Natural philosophers or modern physicists as we call it cannot deny what Galileo discovered nor cannot refute what Newton found out. But when Galileo focused his telescope to the heavens, he did not see God. God never waved his hand and indicated I am here. Even the formulae of Galileo, Newton or Kepler on motions do not need a variable G or God to make the formulae work. It was already assumed that these mathematical formulae do not just explain nature but they are nature themselves. And since it was also assumed that God is the "great mover," He is the cause of everything that moves, then where is He in all these formulae? Since it was assumed that these mathematical formulae are essentially the nature of the inertial phenomena, they seem to work even without God as the cause to make the formulae work. God, the great mover, therefore, seem to be unnecessary in all these mathematical formulations to find out how nature works. The formulae are self-contained and self-operational.

It is for this reason that Descartes had to philosophize at great lengths to establish that God exists before proceeding to his dictum that "he thinks, therefore he is." It is also for this reason that Berkeley also had to contend with the empiricists and materialists of his time, arguing that it would promote the belief that our senses could mislead us to the nature of material things and could skeptically sway us to believe that God is unnecessary to make the material world operate without God's assistance.

But amid the skirmishes of the rationalists and empiricists, it was reason that made its triumph to uncover the operations of nature and to order our social lives. What then, is scientific method really deductive or inductive? The challenge of Bacon and Descartes on Aristotle's syllogistic technique is a problem of ascertaining the truth of universals which stand as the major premise of the process. But the inductive method that Bacon proposed also beg to ascertain the truth of the initial statements that would present themselves for confirmation. Even experimentation which Galileo found so much success needed the guide of a major premise in order to secure the direction of the experiment. In all these instances, theory is at work. The major premise in a syllogistic enterprise is a theory that could find confirmation or falsification in an experiment for an experiment also needs a theory to pave its pathway, even the inductive process that also needs confirmatory or falsificatory experiment is also theoretically guided in terms of the statements that it hopes to prove, affirm or reject.

But despite all the methodological developments, despite the epistemological debate, what role does theory have in all these cases?

No experiment would ever be made without a theory to guide it. It is an extreme impossibility that an experiment would be mounted by simply gazing at the heavens and a sudden puff of thought would emerge that would compel a scientist to embark on an experiment that would yield a pathbreaking result or a stupid one. What would a scientist experiment on if he has no idea of what to work on and how to go about it? What variables would he manipulate in the experiment without the hint that a theory could give? In the same manner what would one induct about without the theoretical propositions? The statements which the inductive method would rely on are actually derived from theoretical formulations. In the same manner, what would one deduct from if there are no general statements aided by theoretical formulations? The deductive method begins with universal statements that theories can provide. What would one deduct from without theoretical formulations? It could be a law. One could deduct from a law since they are irrevocably proven true. But a law originates from theoretical statements proven true by mathematical deduction or experimentation. Since a law is a universally confirmed statement or set of statements, what would be confirmed or falsified without theoretical positions? If these laws yielded mathematical formulations taken not simply as "appearances" but essences of the natural order then what would there be to mathematically formulate without first formulating theoretical statements?

Theory and Science Defined

This abbreviated discussion springboards into what theory and science are. Theory is a textualized system-formulations of the deconstructed, reconstructed and foreconstructed phenomenon. A phenomenon, in this context, can be defined as a regularized or patterned occurrence. A theory is a set of statements meant to formulate concepts that attempt to mirror phenomena that besiege human beings in the natural or social order, where such statements are capable of deconstructing the phenomena into parts, reconstructing or building them back again into a relatedly functioning whole, and capable of providing foreknowledge of how such relatedly functioning whole operates.

A theory is like a diagram or blueprint of an engine. It contains a portion bearing all the named parts and a portion of how they fit in. But the diagram also contains an exploded view of how the parts can be disassembled and be put back in again. A mechanic who has the diagram in his possession does not need to bolt off all the parts anymore to see how the engine looks like inside and out, nor would he need to disassemble all the parts to find out what is wrong if it starts to malfunction. A mechanic could specifically work his way into the engine at a specific location where it seems to operate badly.

There are actually two views of theory: the received view and the semantic view. The development of these two views is closely intertwined with the development of the philosophy of science. The received view posits theory as axioms formulated in a mathematical language. The theory as a set of statements also possesses a dichotomy of theoretical terms which should find direct connection with observational terms mediated together by correspondence rules (Suppe 1999:23). Challenging this position is the semantic view which assumes that theories are language-systems that create mapping relations with phenomenon. It posits that theories belong to a family of models that try to make a picture of the natural order (Suppe 2000:S105). A lengthy discussion of these two views will be dealt with in the succeeding chapters.

Though theory is a text it is different from any other text or literature. The statements in a theory are power-imbued to encapsulate an occurrence in regularity, for it can tear it down in order to reveal its pattern, it can assemble it back again to reveal how it works, and it provides a vision of the results as the phenomenon occurs or limits itself to work.

Theories are not facts. This is contrary to what Prof. Harold Kroto claims. Prof. Kroto, a Nobel laureate in chemistry argues that there are two types of theories: scientific and unscientific. "Scientific theories are considered true or facts because they have been found scientifically to work and we know why they work" (Kroto 2008). Theories are not facts because facts are loose empiricals. Facts are pieces of observables which are still uninterpreted. We interpret these pieces of observables and we actually take the guide of theories to help us interpret them. Theories are the statements that have the ability to organize and make sense of the facts. Facts, therefore, exist independently from theories but it is through these theories that we are able to understand and interpret these facts. The reason why Prof. Kroto placed these two concepts in correspondence is because of the success of Newton's theory of gravity or Maxwell's theory of electromagnetism to account for facts that exist within their limits.

In this case, theories are tools to uncover. Students of automotive mechanic who want to know how a car's engine work would disassemble an engine in order to find out its parts. To accomplish this, the student would need various tools to disassemble the engine. On the process, the student would know the parts of the engine and how its components work.

But since each tool in an automotive shop is limited to the use of a particular job, theories are also limited to explain a specific type of phenomenon.

But other than a tool, theories are apparatuses to organize. After disassembling the engine, the student would then re-assemble the engine and make it work. The tools that aided the student to disassemble and re-assemble the engine are pieces of equipment to help the student understand how the engine works. Theories then work in the same manner as tools do and in the same way as apparatuses to put the parts back again. Theories disassemble a phenomenon into its patterned parts and rework to assemble them back into their working parts. But theories are not just tools and apparatuses in order to understand how a certain phenomenon behaves or operates. If an engine fails or breaks down, a student would already know why it did since he already has a peek at the inner workings of the engine. Without disassembling the engine to find out its trouble, a student would already have a foreknowledge of why it fails. This is also how theories operate like apparatuses and that is to provide foreknowledge of the results or consequences of the phenomenon. It's just like saying, if this phenomenon happens, this will be the result. With the theory of gravity, if we throw a stone above, it would fall back towards the earth. With the theory of special relativity, a matter of certain mass would give off energy equal to its mass and the square of the speed of light. With the theory of quantum mechanics, a particle's location cannot be known if its velocity is determined. With the theory of communicative rationality, a transaction possessing, truth, sincerity, and right speech would result in agreement. With the theory of absolute advantage, a nation will export a commodity which it has the maximum advantage or the utmost productivity in producing the commodity given two products to choose. With the theory of pure competition, prices among products that are pure substitutes of each other are kept at the same level when multiple firms which have the freedom to enter and exit the market produce these homogenous products.

But just like any tool, a theory has its limits. Between an open spanner and pliers, both can turn bolts and nuts but an open spanner has a certain limited function to turn bolts and nuts just as pliers do. Two theories can account for the same phenomenon but one theory can explain the occurrence in a different dimension as against the other theory. Beyond these limits, the theory already fails to explain.

But other than tools and apparatuses, theories have enabled us to "read" the phenomenon. The process has transformed the phenomenon into something that could be perused. Even if the theory churns a formula, the mathematical statement is still a text and it is with this text that we are able to read and manipulate the world imbued with various phenomena. Our reading and manipulation of the world through theory has proceeded with textualization. And theorizing is textualization since a theory is a text. Textualization is a bounded enterprise of rigor where the reality that the enterprise seeks, the method to seek for the reality and the rules that govern both the seeking and the subject of what is being sought are all proposed, validated and challenged in constructs that encapsulate the phenomenon as a subject that is meant to be read and comprehended in a process that deconstructs, reconstructs and foreconstructs. This definition could have aimed at seeking for truth and not just reality but truth is the ultimate where no amount of disputation could overturn its truthfulness. Realities are interpretations of what is observable. Since science is more concerned with observables, then reality is the object of theory.

Science is rigor-bound, method-driven, textualization of patterns of regularities and irregularities in phenomena that exist in nature or in human relations through deconstruction, reconstruction and foreconstruction in an elegantly convincing fashion. Science is not just nature or human relations textualized, it is also the method and validation textualized. Science then is a body of rigor-achieved, rigor-differentiated and rigor-validated text that can create an image of the world it attempts to explain or understand; represent or model; discover its order or even re-order what it has discovered. Science is not just about the subject but also about validating how we have come to know what the subject is. Science then is epistemology at its best.

On the other hand, science is all about the human search for how the huge mechanism of nature and how the huge engine of human relations work. The rules that govern the search and the rules that run the validation of what was searched are all human inventions. Science then is a human invention. And it is through theories that humans invent that make for science. Science is not about how God created nature or how evolution took its course to produce what is in nature. God may have named what is in nature differently from what we have ascribed it to be. God may have looked at how these things worked differently from how humans see it work. Science is a human enterprise inasmuch as it is a human invention to unlock how creation works or how evolution proceeds.

Science proceeds with textualization that theory has enabled humans to read and manipulate the world. It is this enterprise that has personified reason... for it is reason that has gifted humans to overcome the world.

Chapter 3
A Milieu too Many

The short interlude after the Renaissance propped up the victories of reason in its revolt against the Catholic's monopoly of the acceptable things that we know. From Galileo's challenge against the Catholic's belief of an earth-centered universe to the empiricists-rationalists debates of Locke and Descartes to name a few, individuals who adhered to reason were called enlightened ones. Even Catherine the Great of Russia, who devoted a great deal of time conversing with the French liberal Diderot, was also ranked with the enlightened individuals, while Diderot though a rich man came to the tsarina for financial rescue (Copleston 1960: 41).

The Age of Enlightenment heralded the split of advancing knowledge towards discovering the natural order on the one hand and reorganizing the social order on the other. The end of the 30 years war in 1648 was a trigger to this era, when after the bloody war between the Protestants and the Catholics, thinkers began to disaffect themselves from the sway of religion towards knowledge. This was a glorious revival of the Greek classics with which Plato and Aristotle used reason to rationalize the best political organization. True enough of the split, the separation became more contentious as the geographical divide along the English Channel of England and France became obvious with the directions of their ideas.

The English had equal preoccupation with the natural science and social philosophy while the French tilted more on social ordering (Copleston 160:121). The English enlightenment had its love affair with Newtonian science. Reason was the champion toward the discovery of the world. The French Enlightenment on the other hand had its romance with Rousseau who was trying to use reason in order to find the best social ordering with the advancement of the general will. For the French reason was to serve the society by founding its best order. Enlightenment braved the cudgels of denouncing the Catholic Church which murdered reason and advanced social contract and liberalism as the foundation of government system. Reason was hailed as the hero that ended the long intermission of science's stunted growth during the medieval age. As science progressed in the natural and social setting, competing claims were advanced purporting to breakthrough the natural and social order into concepts that theories forward. The triumph of these theories is also the victory of reason.

> *"In the beginning was logos... that logos was reason... that reason was in the text... and that text was theory."*

If theory could speak, it may have uttered, "the beautiful line in my story started with the thought that the world can be understood." But why care about understanding the world anyway. We go through the scheduled and unscheduled motions of life without thinking about the universe or the social system where we belong. You might think that these things are only reserved for those who are theoretical fanatics. But the world is a theory's preoccupation. It is its life. It is its passion.

Theories would perhaps have been born amid sleepless nights and bouts of coffee brew. When reading a novel or watching a movie, you come home discussing the story as if it is a real life story in real time. But most of it is simply fiction. Theories are not fiction. You talk about theories. You attack theories. You teach theories. You try to debunk theories. Then you use theories. Theories are alive even if those conditions for which they were conceived do not exist anymore for the debates, discourses and references resuscitate them beyond the deathbed of the bookshelf. Theories don't just live in mental abstractions, or on the yellowing pages of the book, but theories exist through debates, discourses and interactions.

The problem with condemning theories as simply abstractions comes from the idea that what is real is observable. The empiricist hegemony that knowledge emanates from sense impressions would be shaken if challenged that not all that is real could be directly observed. The atoms with which any matter is constructed cannot be observed. No one will ever see an atom for an atom is smaller than the smallest quantity (quantum or photon) of light. If an atom is that minute, how much more the protons and electrons? But atoms and their sub-atomic compositions have been calculated to be real. Their effects (and not they exactly) have been observed.

Neptune was discovered not because of its physical traces passing through the lens of a telescope but through the aberration on the orbit of its neighboring planet Uranus. William Herschel discovered Uranus in 1781 but the irregularity in its orbit had been an astronomical puzzle among astronomers since it defied Newton's Law of Gravity. In 1843, an English astronomer John Couch Adams, fresh from his graduate studies, started plotting the traces of Uranus's path and predicted that another undiscovered planet was causing the discrepancy in Uranus's orbit. In October 1845 Adams finished his calculations and submitted them to other astronomers only to be shunned. Meanwhile, the French astronomer Urbain Jean Leverrier theorized on the same problematique and sent his calculations to Johanne Galle of the Berlin Observatory in September 1846. Galle went to work on his telescope and in the evening of September 1946, he found Neptune, only 2° short of Adam's predicted position (Seeds 1988, 526). What brought this to the open? How did they know these things before hand? The answer is... theory. Human beings have brought theories to life and they are sustaining the life of theories through their interactions.

Through theories the world can be explained. The world can be organized. The world can be mastered. But what is the world in the first place?

The world can be understood in different viewpoints. Human beings are familiar with separating their social world from that of the physical world. That is fair enough. Human beings anyway are different from the environment where they live in. In this view, we are entities placed in a physical habitat. There is the physical world serving as the dough and the social world acting like toppings. There is the universe and we human beings are in it. But there can be another way of looking at the world.

The three milieus

The world can be viewed in three. The **milieu-already-constructed** is the milieu already made. It is the milieu that is already there even before it impinges on the observer. It is the milieu already in existence. The universe is an example. It is something, which human beings never constructed. But it is already there. Since this milieu is already constructed, what the observer does is to discover it in order to uncover its nature and its dynamics. The idea is to figure out the principles on how it works. And since such principles are generalized and can apply to all phenomena of the same nature, then such principles become a "law."

Don't ever think that this refers only to the physical world. The social world of human beings can also be seen as a milieu already made. For human beings, this is a milieu that has already transpired or is transpiring according to some sets of principles. These principles are like gullies where water flows. Unless there are other routes, water will naturally flow on those canals. The principles are like these gullies. And these principles can be generalized. This milieu is already constructed because of the principles at work. In all cases, this is a predictable milieu, as long as the principles have already been discovered. This milieu is highly regular and patterned. Insufficient principles that could not account for some phenomena could be termed an anomaly which could be resolved if another principle could be discovered to account for the regularities occurring along the thread of such anomalies.

Then there is also the ***milieu-under-construction***. If the milieu-already-constructed appeals more to the physical world, the world-under-construction finds more affinity to the social world. This milieu considers human beings as active participants in this milieu, which they build and change through their interactions. This is a milieu that is never complete and will never be, since human beings are prone to change or to reconstruct it. Human beings may not know it, but their interactions continually produce and reproduce their milieu. Their milieu may have patterns, but such patterns may be obsolete. Another pattern at some point in time is recreated. What an observer does with this milieu is not to discover, but to interpret. The idea is not to discover principles, for such discovery may not have regularity across time and space, but the objective is to uncover this milieu, and in the process of uncovering, an interpretation is made.

If principles are the tasks of discovery in progress, it is meaning which interpretation hopes to derive. Meanings are not like gullies where water flows. They are like the humans themselves wading on those gullies, who may wish to go to any direction they like. Principles create the box, but meanings provide the various trappings.

Nevertheless the physical world can also be viewed as a milieu-under-construction. We are capable of changing our physical habitat consciously or unconsciously through the same acts directly or indirectly affecting the environment. You can change the structure of the landscape and recreate torn out structures, even present your bodies for bacteria and virus to mutate. The universe is expanding and stars die and others are created. But the challenge which quantum mechanics poses on how we view the world also changes the way the physical universe can be conceived. The traditional view that the world can exist even without consciousness is shattered by the thought that since the observer can affect the motion of particles, the physical world would not have been without a consciousness.

It is these milieus then that theories are concerned with. This does not mean, however, that there is a complete separation of these two milieus. Plato, a Greek philosopher of rich imagination, devised an analogy to depict human's search for knowledge. In a teaching moment with their teacher Socrates, Plato recorded in his bookd the *Republic* of their class discussion, which today is aptly called the "Allegory of the Cave" (Kreis 2012; Sparknotes 2015). This time, make your own imagination work.

There is a cave that descends deep in the earth with two men shackled facing a wall. They were chained there all their lives and could not look back towards the entrance of the cave. Several objects pass by the entrance which cast shadows on the wall. Since these two individuals had only seen these moving shadows all their lives, they would think that these shadows are the real objects. Socrates goes on to ask, "what if one fellow was able to get out of his chains and was able to emerge out of the cave?" At first glance of the light at the entrance, he would have had difficulty looking at the glare of light he had seen for the first time. If he saw the same objects pass by, then he would realize for the first time that what he was seeing all along on the cave's wall were simply shadows. He wouldn't have believed it at first. But upon close examination, he would have had associated the shadows with the real objects he was observing. At this point, he had a confrontation with the natural world. This is the "milieu-already-constructed" but at that point, he was constructing it on his own. Socrates goes on by posing, "what if he goes back to his friend and tells his discovery?" Either his friend would believe him or not. What is important is that, at the time of their interaction, they were both constructing their milieu. Thus we have a "milieu-under-construction."

My point is, these two milieus are never separate from each other. There is the milieu-already-constructed, a milieu which needs to be discovered, a milieu out there which will perpetuate even without human beings to tinker on. But that's not the end. This milieu has impinged on us because of our construction of it. This is the milieu-under-construction. Our interaction, processes of interpretation and discovery lead us to build this already constructed milieu into a milieu we construct for our own. Simply put, this is science in our language. Science therefore is nothing but the meeting of these two milieus. Science is a human invention. It is the result of a constructed milieu based on the empirical outlines of the milieu-already-constructed. But that is not the end of it. As theories attempt to mirror the milieu, theories however, shape the way we configure it. Since theories are a construction or reconstruction of the milieu, theories on the other hand, construct or reconstruct our reality. What then is the theories' role in these two milieus? If science is a human invention, theories are also human inventions. Human beings have created them.

The amazing thing about this, in fact, is that though theories are born in a milieu-under-construction, the subject matter may be about the milieu-already-constructed or the milieu-under-construction itself. Theories therefore bridge the milieu of discovery and the milieu of interpretation. Theories are, in fact, an interpretation of these two milieus. But don't be mistaken that theories bring about the milieu-already-constructed. It is already built. Theories do not create that milieu. That milieu exists and perpetuates even without theories or even if theories don't faithfully reflect that milieu.

But when we become confronted with how this milieu was constructed, we begin to inquire, interact and textualize how this milieu has become, thereby creating the milieu-under-construction and bringing theories about. The inquiry, interaction and textualization never stop. Theories are brought into being, even if they have been a previous mistake or an inconsistent. In this case, a theory is revised or another competing theory is created to attack and debunk the former one. Still theories are being interacted with. But either way, in a theory's lifetime, being supported or attacked blows the breath of life on a theory and keeps it alive.

But amazingly still, we are not satisfied in finding out how the milieu-already-constructed has been made and sustained. We are also intrigued on how our interactions take place. We are concerned with how the milieu being constructed occurred. Thus another theory is again created to account for it. Theories are human inventions. They are a text. Through them the milieu already constructed and the milieu we construct can be read.

Is a theory therefore, the milieu that impinge on our senses? If you analyze the theory of gravity for example, and witness the spoon fall on the floor, you did not see a theory.

That is the milieu-already-constructed at work. When you consult the Theory of Gravity, read about how it accounts for the falling of the spoon on the floor, that is the theory at work. If you witness a group of people interacting by way of speech and gestures and there construct their reality on how to deal with certain issues, you did not witness a theory. If you read about Symbolic Interactionism, a theory` that could explain such an occurrence, then you have a theory at work.

Nature and human relations live and operate in these two milieus: the milieu-already-constructed or the milieu-under-construction. A theory is a text. What a theory does is to organize, through texts, patterns in order to give account of the phenomenon operating in these two milieus. What a theory does is to pull some principal threads or patterns of relationships at work and weave out an overall view of how these patterns function. Theories are made of logically fitted symbols with precisely designated vocabularies possessed with their own values and meanings.

We have to bear in mind that science is a human invention. Human beings have assumed that science is out there concrete and not affected by the subjectivity of their practitioners. It is true that if an astronomer gazes at the sky through his telescope and theorizes about the exploding stars, he is investigating something concrete. But how the pieces of what he knows are systematized in a body of knowledge and how this embodiment and exposition are accepted in a community of scientists make up what science is all about. The milieu that a scientist is theorizing is out there but that is not yet science. Once the scientist conceptualizes, systematizes, and exposes his knowledge in a market of ideas where it is challenged and falsified, then what you have is science. Thus science is a human enterprise. It is a human invention. Even if scientists talk of the natural world, they are still rationalizing a world in the eyes of human beings. The only difference is that they have a method. This is not the world in the eyes of God. If you ask God about a natural phenomenon and He answers you, He may use different terminologies and may even explain it in a different way.

Thus, as human beings are imperfect, science which they invent is diluted with imperfection. This is the reason why the whole practice of science carries with it the need to challenge, confirm or falsify claims. Yes, once human beings hark on the word science, they clarion that science is divorced from the biases of the practitioner. But in practice, a claim in science is presumed to be correct if many scientists agree with such a claim. Remember, practitioners of science are limited in their view of the world. While truth and validity are not determined by how many scientists agree with the claim, the claim, however, is adjudged to be correct by how many illustrious practitioners agree with the claim, only to be dumbfounded in the future that what they agreed on is not true or deficient of truth. This whole endeavor, therefore, is a human invention.

Scientists discover and interpret. They discover the world and conceptualize on it. Then they invent. They invent terminologies to account for certain concepts. They invent labels. They try to interpret how these concepts relate with each other. They invent a system of how these concepts are interpreted. They invent how these systematized concepts can be validated. In short, they invent a text. They invent theories. They invent theories with a goal to capture and reflect the world. Theories are therefore a system of language designed, with evidence, to mirror the intricacies and dynamics of the world. This last statement is very significant.

As a system of language, theories are made up of terms precisely defined to capture concepts. These concepts are units which theories analyze. For this purpose, theories could be likened to a virus which has a designated zone or part of the body to attack. The terms with which theories are made up of are precisely defined to analyze a unit in order to bring about a concept. These terms and concepts are linked together by virtue of certain relationships. Here is a system made. This is the ***milieu-of-the-text***.

For example, Isaac Newton's Second Law of Motion states that the body in motion will remain in motion unless force sustains it in its movement. The force (F) needed to sustain its motion is directly proportional to the mass (m) of the body and to the rate of change of the body's velocity or its acceleration (a). If you roll a ball on the floor, it stops at a certain distance. To keep it moving, you would need to continuously apply certain amount of force. Here we have a concept of force as constituted in Newton's Second Law of Motion. We have terms defining certain concepts such as mass and acceleration. Independently, mass is the amount of matter in an object while acceleration is the change in speed at a certain time. But if we put them together in regard to certain relationships then we have a concept of force where the greater the mass of the object, the more force is needed. The higher the mass however, the slower the acceleration.

Here is a system of ideas made. Theories are made up of units of analysis which are linked with each other via certain relationships. This system of ideas is nothing but language. Moreover, language devised to explain the milieu-already-constructed is simply a medium to reflect the milieu it seeks to explain. The terms are ascribed by the scientist or theorist. Thus came the project of the Vienna School on the purification of language to correspond with observable reality (Outhwaite 1983:7). This language serves as the medium for theory to use. But for the milieu-under-construction, language is the theory in itself. The theorist captures the terms which the subjects use. The terms are not implants by the theorist but they are the actual language which the subjects utilize in their everyday lives. This is theory on the ground or grounded theory.

But through language, theories have been created a text. Theories are texts which make the world readable. They are a system of ideas that translate the world into text. They are language systems that mirror the world in order to construct or reconstruct reality. We build theories to reflect and interpret the world but theories shape the way we configure or make sense of it. This is constructed or reconstructed reality for us. We make theories but theories make us as well. These are the main arguments of this book. It is how we see the world and read it that we have come to change it. The next few chapters will illustrate these arguments through the two milieus I am talking about. But then if theories are a text and they mirror the world, the next question would be how faithful are they in reflecting it? The succeeding chapter will answer this question. In the process, I will advance the "Textualized Nature of Theories." But you have to remember that theories are constructed not to seek for truth. Human beings who constructed theories have only uncovered or discovered a portion of what we may call truth, which becomes tainted with color, and relational subjectivity once relayed and becomes a subject of human interaction. Sometimes we loosely talk of truth. But it is not objective truth that we seek. It is constructed reality.

Journalism, which is a very unscientific preoccupation, prides itself as one which digs for truth. But when journalists go out in the field and gather facts, observe events, ask for interviews, conceive a slant, write the news, and the editor edits it with his own slant, what was produced is not truth but a story, based on how they conceived the story to be. What they collaborated on, though based on observable facts, is a construction or reconstruction of reality. It is not truth but a reality constructed based on how they conceived it to be or based on how it impinged on their senses and consciousness. It is constructed or reconstructed reality based on how they think or sense it to be.

Now take a court trial as a more tedious process than journalism. Here human beings have an equal opportunity to present their own sides of a single story unlike journalism which does not have an institutionalized procedure to present two sides of adverse positions. Now lawyers here do not just speak simply of facts, they talk of pieces of evidence that have undergone strict scrutiny. In the environment of questioning, pleading, manifestations, stipulations, where speech and reason are the weapons, judgment is rendered not on the truth which they pride about they seek. But what the two confronting lawyers in front of the judge and jury have created and have tried to confirm is a constructed or reconstructed reality. It is constructed or reconstructed reality based on how they disputed it to be.

Theories are constructions and reconstructions of reality. More importantly, this construction or reconstruction works. It is this construction and reconstruction that make science a human invention. It is an invention that is made up of rules. It is an invention created in language. And such language is also a human invention. It is an invention that lives in the text. And the text is a human invention. Science, then, is an invention that necessitates the invention of theory to construct or reconstruct reality. It is this construction or reconstruction that has translated the world into a text. And through this text, we have come to understand and master the world. Thus it is the way we see the world and read it, that we have come to change it.

Can theory be constructed without textualization?

No.

Can science proceed without the text?

Never.

Chapter 4
The Milieu-already-constructed

The milieu-already-constructed is a milieu in pre-patterned state. It is the milieu that is already there. It is a milieu that is fixed and characterized with certainty. It persists outside the observer and perpetuates even without an observer. What textualization does is to disassemble this milieu into parts, represent them into terms that would express concepts and see how these concepts work by virtue of their relations and mechanics, then reconstruct them back to see its wholeness. Since this milieu is already constructed, an observer, theorist or scientist is there to see how it is made. How it works.

Textualizing Nature's Mechanics

Aristotle defined nature as a principle of motion and change and that motion is the fulfillment of what exists potentially (Hutchins 1952b: 278; Aristotle "Physics" 201). With the triumph of Newtonian physics in explaining the natural realm, physics was more viewed as the science of motion. In here mechanics will be explained by two theories within the pale of Newtonian physics and Einsteinian mechanics.

Let's say, you are riding on an air-conditioned bus. It is cruising along at constant speed. Now as you fix your eyes on the headrest of the seat in front of you, you noticed a fly resting on its velvet covering. Think for a moment. What if this fly suddenly wags its wings and flies? Well, since every time the bus jolts forward, you are yanked backward towards your seat then, common sense would tell you that since the bus is speeding forward, the fly, once it releases itself from the headrest, would be left behind splattered on the window at the back of the bus. But would it?

Nature dictates, that is not what happens. As the fly releases itself from the headrest, it could circle around, could fly above the head of the person in front, could soar to another person, and do anything it wants. It would never get splattered on the window at the back of the bus unless the conductor's fly swatter could catch it on the windshield and it ends its journey right there splattered in front of the driver.

The bus's motion can easily be explained through Newton's First Law of Motion which states that a body at rest remains at rest unless subjected to an external force or remains in constant motion unless an external force is applied to halt it. This means that a stationary object will remain at rest unless force is applied to make it move and a body in constant motion will not stop unless force is applied to make it stop.

The speed (S) of the bus is equal to the distance (D) the bus covered and the time (t) it covered that distance (Figure 4.1). Since the bus is traveling at a constant speed then speed can be measured by dividing distance over time.

$$S = D / t$$

The Milieu-already-constructed

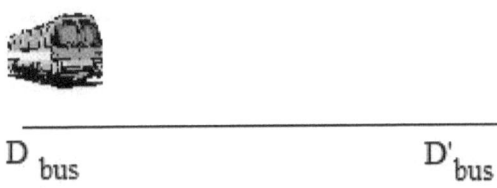

D_bus   D'_bus

Figure 4.1 Bus Moving from D$_{bus}$ to D'$_{bus}$

But since the First Law speaks of force, the Second Law states that the acceleration (a) of a body in motion is directly proportional to such force (F) but inversely proportional to its mass (m).

$$a = v/t$$

$$F = m(v/t)$$

$$F = ma$$

The greater the mass which is an obstructing component, the greater the force needed to accelerate it. The higher the rate of change in its velocity, the greater the force needed. Thus if the bus starts off at rest, the driver would have to step on the gas harder at low gear to apply more force for the bus to move and the harder he would press on the gas if he would like to accelerate the bus faster at lesser time. The change in velocity over a period of time would therefore need more force. But as soon as the bus coasts along at constant speed, force turns zero since there is no more change in speed needed, thus the driver maintains his pressure on the gas.

But notice that as the bus accelerates forward, you are pressed on your seat in the opposite direction. As the bus brakes and slows down, you are driven to lean forward.

Newton's Third of Law of Motion explains this. It says that for every force that acts on the body (acting force), there is another force that acts upon the body (reacting force) equal in magnitude but opposite in direction.

$$F = -F$$

$$m_1 a_1 = - m_2 a_2$$

But as the bus cruises along at constant speed, you will find that you can walk to and fro inside the bus inasmuch as the fly can hover anywhere inside in.

Newton's calculations have several consequences. As the fly desires to soar forward, we can calculate its speed by dividing the distance it has covered from the position of the headrest to the windshield in front over the time it has taken the insect to fly that distance. Here Newton's formula still holds. But that only holds because the fly views the inside of the bus and all of the passengers in it to be in a state of rest.

If you open the window and the fly flies out, quickly, it would experience how it is like for the bus to travel at let's say 100 kilometers per hour and be left behind instantly at 27.78 meters during the initial second disregarding yet the wind that sweeps it farther behind. For the mosquito and the passengers inside the bus, it seems there are two frames of experiences. There is the experience inside the bus in relation to the passengers sitting inside it, which can be perceived to be in a state of rest. But if you look through the window, there is another frame of experience, which you see to be moving, but it is actually the bus in motion. This has great consequences for Newtonian theory of motion. You will find that there is no absolute standard of rest.

For example, a 100-meter sprinter started off at the sound of a gun and breasted the tape at certain duration. His

The Milieu-already-constructed

speed would be easy to calculate by simply dividing 100 meters over the number of seconds the clock ticked. But that could be done since you presume that the track is not moving. We view the whole athletic field itself to be at rest and only the runners are the ones in motion. Here Newton's calculations hold.

The mosquito that flies in mid-air stays freely in mid-air inside the bus. The runner on the track perceives that the athletic field to be at rest but the earth rotates and orbits around the sun. Thus the track which is at rest is relative to the earth which is in motion. If you decide to zoom out in outer space, then you will experience the earth's spin just like the fly that was swept back once it flies out of the bus's open window.

Story has it that Galileo Galilee climbed atop the leaning Tower of Pisa to test if two objects of different mass would fall at the same time. He did the experiment by holding a piece of rock and a piece of paper crumpled like a ball, and throwing it at the same time. Galileo found out that no matter what the mass is, the two bodies would hit the ground at the same time. The reason? Galileo found out that the acceleration of a free falling body is constant at 9.8 m/sec^2. What if someone did the same experiment but this time with certain variation. What if he throws the ball and the piece of paper and in the process jumped out of the tower together with the two objects? As he looks at his surroundings, he will experience his fall but as he looks at the rock and crumpled piece of paper which are falling together with him, he will find himself as if suspended in mid-air.

The first thesis of Einstein's Special Theory of Relativity accounts for this observation. The analytical form of physical law is the same in an inertial frame of reference. One inertial frame moves. The bus for example. You could measure the speed on how the mosquito flew from one point to another inside the bus. At a uniform speed you would experience the bus to be at rest. That is one inertial frame. If you are an individual outside the bus, you would see the bus uniformly traveling and experience the ground you are standing on to be at rest. That is another inertial frame. In each inertial frame of reference, the physical laws are the same.

As matter is geometry in space, matter is confined in dimensions. Dimension here means space or confines where matter can move. A space without dimension is a point. No object can move at a point except at that exact point. That means the object stays where it is, unless, of course, the object is a lot tinier than the point where the object could squeeze itself and move inside it.

A line, however, is a location with only one dimension. That means, the object in the line can move either horizontally (left to right or vice versa) or vertically (upward or downward) depending on how the line is drawn. The object can even travel diagonally if the line is drawn in the same way. The characteristic of the object in this linear one-dimensional location would also look the same as to how the line is drawn. A crooked line would make the object crooked and make it travel in a crooked path. The same is true with a wavy line or a spiky one. In this one-dimensional location, moreover, the object would have neither thickness nor width – only length. Thus it could be measured only by virtue of its length, going left and right or up and down.

A two-dimensional space is a plane. By two dimensions, it means that an object can move either left or right, up or down, in whatever direction the object would be compelled to travel or desire to move. The object now has two directional options, horizontal or vertical. This is a space were objects can have both length and width but never thickness. Thus the two-dimensional space can be measured by virtue of its area (length multiplied by its width). A piece of paper is a

The Milieu-already-constructed

two-dimensional space. If we draw and cut out a person and place him in this space, he would travel in the direction of the space's length or width but never beyond the edge of the paper lest he falls off.

If you look back at the bus which is traveling uniformly along a straight line, the displacement of the bus could be plotted along the X axis from point X_1 to X_2 at velocity V and at time t. This is one inertial frame from your observation while looking at the bus from the outside (Figure 4.2).

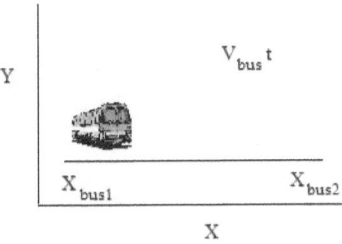

Figure 4.2 The Distance Traveled by the Bus

Now while inside the bus a mosquito lands on top of the headrest of a passenger seat in the middle of the bus. Since the mosquito is simply perched atop the headrest, the mosquito is at rest. This is another inertial frame from your observation inside the bus. If the mosquito flies forward, it has covered another measure of distance from X_1 to X_2 at its velocity but with the condition that the time both actions occurred for both the mosquito and the bus is the same (Figure 4.3).

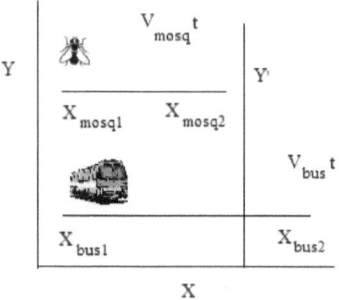

Figure 4.3 The Mosquito and the Bus at Two Inertial Frames

The bus reached point X_{bus2} due to the velocity V_{bus} which it covered the distance from X_{bus1} to X_{bus2} at time t.

$$V_{bus}t = X_{bus2} - X_{bus1}$$

$$V_{bus}t = X_{bus}$$

The mosquito, on the other hand, flew at a distance X_{mosq} with a velocity V_{mosq} at time t.

$$V_{mosq}t = X_{mosq2} - X_{mossq1}$$

$$V_{mosq}t = X_{mosq}$$

At this point, you would realize that there is no such thing as absolute rest. A body at rest is relative to another body in motion. But at this stage, the discussion simply revolves around a flat representation. The world where you live is a world not simply limited to an X and Y coordinate. You live in a world of more than 2 coordinates.

The Milieu-already-constructed

A cube is a three-dimensional world. By three dimensions, it means that an object can travel left or right, up or down, forward or backward. There would be three directions in his movement and this world does not only have length and width but also thickness. It is measured through its volume where length is multiplied by its width and its thickness. You live in this three-dimensional world. Thus everything around you is seen with length, width, and thickness. Even a piece of paper once viewed on the side will be viewed with certain "thinness" and will be made measurable through a system of "micro" measurement. A point will be gauged by a three-dimensional being with volume since he could imagine shrinking himself smaller than the point and move inside it. And that is how we human beings see things.

Yet a three-dimensional being can do wonders on a two-dimensional object. A three-dimensional being can strip off a flatlander from his flatworld, vanish him in his two-dimensional plane, and transport him in another flatworld. What can confine a flatlander is simply a line drawn around him but he could easily escape his confinement if someone in three dimensions takes him out of his world, momentarily swifts him in the three-dimensional world and lands him in another area in the flatworld. A flatlander would also be startled by his new experiences if he would be delivered in a three-dimensional plot.

Imagine if the flatworld or the plane where flatlander dwells is folded into a cube (Figure 4.4). You, in three-dimensions, can easily see all the sides of the cube even imagine the empty space inside it. But for the flatlander who was not able to escape, he would have different experiences at different sides of the cube. Since he is a flat inhabitant, the space inside or outside the cube is unknown to him for he could not go beyond it lest he drifts. When he goes near the edge of the cube, he would find himself being bent or warped until he arrives at the adjacent side.

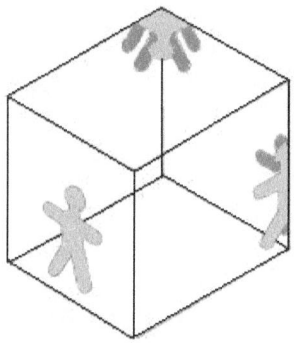

Figure 4.4 A Flatlander on the Surface of a Cube

If you paint the sides of a cube with different colors, once in yellow, the flat inhabitant would be in red at an instant, travel farther and arrive in blue, move on until he goes back to yellow again, the place where he started once. The same is true if he travels vertically. He would be at different places and would land at the same spot, witnessing various colors at different events. If he stops at the edge, however, he could see two colors in separation while he himself is bent. If he halts at the corner he would witness three separate hues at the same time while he is bent in three faces as well. He could conclude that he is just going around from where he started. Yet he would find six different colors representing six different sides. But it would be difficult for him to conclude he is in a cube for a cube is unknown to him since everything from his vantage point is flat. But you, in three-dimensional world, could witness how it all happened because you either move to the left or to the right, jump up or down, and step forward or backward. Our biological system is confined in these three-dimensional directions that we experience and

The Milieu-already-constructed

interpret space and the objects around us in three-dimensional plot (Figure 4.5).

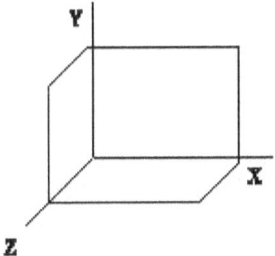

Figure 4.5 A Three-Dimensional Plot

Going back to the bus now, the bus travels diagonally across X,Y,Z then the generalization of its line covered would be achieved through the Pythagorean formula (Figure 4.6).

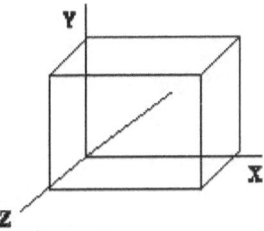

Figure 4.6 Line that Travels along Three Dimensions

$$V_{bus}t = ((X_{bus2} - X_{bus1})^2 + (Y_{bus2} - Y_{bus1})^2 + (Z_{bus2} - Z_{bus1})^2)^{1/2}$$

If you consider the displacement of the mosquito as it flies across its X,Y,Z axes, the resulting line it traveled would be computed in the same manner.

$$V_{mosq}t = ((X_{mosq2} - X_{mosq1})^2 + (Y_{mosq2} - Y_{mosq1})^2 + (Z_{mosq2} - Z_{mosq1})^2)^{1/2}$$

Previous illustrations will lead you to realize that speed is relative. The speed of one body is dependent upon another considering the frame of reference an observer uses. The dire consequences of these experiences would result in an experience where the absolute standard of rest is shattered. The absence of absolute rest means the non-existence of absolute space. This would mean everything is in motion and one's speed is dependent upon another. But with the relativity of speed, there ought to be something in the universe which travels at an absolute speed. The second thesis of the Special Theory of Relativity as proposed by Einstein states that the speed of light is constant even if the source is moving. The speed of light therefore becomes the reference of all speeds since it is the only absolute speed.

In 1885, James Maxwell was able to unify the partial theories of electricity and magnetism into one force – electromagnetic. Maxwell's equations predicted that wavelike disturbances can be produced by electromagnetic vibrations. Like ripples in a pond, the electromagnetic vibration which is a meter long from crest to trough is called radio wave. Shorter electromagnetic wavelengths are called microwaves while visible light has a wavelength of forty to eighty millions in a centimeter (Hawking 1988: 18). What more, Maxwell predicted the these waves travel at constant speed and if light behaves like waves then light would travel at a fixed speed just like radio waves.

Since then it was suggested that light travels in ether, like sound that resonate on air or like waves that roll on the water. The suggestion of ether was an offshoot of the prevailing consequences of the Newtonian concept where speed is related to something in absolute rest. For light to travel in fixed speed, then it has to be measured to something in relation to. Since ether can be presumed not to inherently move, then light which travels on it and from which light can be calculated relative to would have a fixed speed. Only if the ether will be disturbed will the speed of light be affected as well.

In 1887, Michael Michelson and Edward Morley set up an experiment to test this assumption. With reflectors arranged at right angle and opposite to the direction of the earth's spin, they could observe light traveling at right angle to the rotation of the earth, light reflected back, light parallel to the spin of the earth and light bounced back. Since the rotation of the earth is assumed to disturb the ether, then the rotation of the earth at 30 kilometers per second can also affect the light reflected through, at right angle, and bounced back. Since the apparatus would make the light to converge in one spot the difference in the speed of reflected rays would produce bright and faint reflections. But to Michelson and Morley's surprise, there were no such differences in reflection (Gamow 1961: 93-98). This means that ether has no effect on the speed of light even if the suggested ether is disturbed. Light travels at a fixed speed of 186,000 miles per second.

From 1885 to 1905, several scientists ventured on to study the effect of ether on time which seems to slow down and on shape which seems to be distorted as a body moves through it. In 1905, this belief was shattered when Einstein who was working as a clerk in a Swiss patent office wrote a scientific paper proposing that the idea of ether was unnecessary as long as the idea of absolute time is dumped. Since rest is dependent upon any body in motion, then time is also relative to any moving body. If absolute space does not exist and time is relative, then length as a component of space will also be relative in accordance with how fast the object is moving. And without absolute speed, space and time, and mass which a body contains will also be relative. Einstein's theory results in some amazing consequences.

Newton's equations put forth that speed is determined only by space and time without any effect on its mass and without any change on its length. Conversely, Einstein's time ticks slower in a moving body as it shrinks since space is not absolute. And as it moves, it accumulates mass. According to the theory, this phenomenon can be experienced at speeds nearing the speed of light. And since the speed of light is unmatchable, as the body closes in the speed of light, it shrinks and gets heavy until it reaches the point that it has shrunk and has become heavy enough that it slows down (Hawking 1988: 21).

A moving body therefore contracts in length in the direction of motion. A bus traveling at a certain speed gets smaller in length. If you measure the bus from the front to the rear, theoretically no change in length occurs because you are in one moving body and the measuring device you are using also contracts just as the bus did. The contraction is related to its motion relative to the speed of light.

$$L' = L\,((1-V^2/C^2))^{1/2}$$

But hardly is its contraction felt because the ratio $(1-V^2/C^2)^{1/2}$ of the speed of the bus compared to the speed of light is too small. The same would be true to time. If length contracts in the direction of motion, time slows down as the body moves relative to the speed of light.

The Milieu-already-constructed

$$t' = t/((1-V^2/C^2))^{1/2}$$

And as the body moves, mass accumulates as well. The object becomes heavier relative to the speed of light. Again the phenomenon is not felt because the increase in mass is so small, for the speed likewise is too small compared to the speed of light.

$$m' = m/((1-V^2/C^2))^{1/2}$$

For Newton, space and time are two separate units. But time relatively progresses depending on the speed of a moving body in space. The greater the speed, the slower time ticks while the body becomes heavier as it accumulates mass. A person inside a speeding car would not experience the effect of this phenomenon since the speed of 65 miles per hour (0.018 miles per second) is just a small fraction of speed compared to the speed of light. Time slow down, size shrinkage, and mass accumulation would differ only by a factor of 0.9999999999999998. Too small to be felt. If the car is 7 feet long (0.0013 miles) it would shrink by 6.9999999999999986 feet, a difference of only 0.0000000000000014 foot. A duration of 30 seconds would be 30.000000000000006 seconds in the car, where the driver's wristwatch would tick longer by 0.000000000000006 second and if the weight of the vehicle is one ton or 454.5 pounds, then it has become heavier by 454.50000000000009099 pounds. It only added 0.00000000000009099 pound to its mass.

But suppose he is in a spacecraft traveling at 20 percent the speed of light that is at 37,200 miles per second. Time, length, mass would differ by a factor of 0.9797958. If the spacecraft stands at 363 feet (0.06874 mile) from the nose to the nozzle, then it would contract at 353.66 feet or 0.0673511 mile, 30 seconds would be 30.6186 seconds inside the spacecraft, and if the spacecraft weighs in at 6,500,000 pounds, then it would weigh heavier at 6,634,035.3 pounds at that speed. If a person weighs 120 pounds, then he would get heavy at 122.47 pounds if he were a passenger inside the spacecraft. Yet the person would not have felt time has passed by slower and his mass has increased since his psychological mechanisms has also slowed down, thinking everything is still normal. The space shrinkage, time slow down, and mass accumulation, therefore would all depend upon the factor of the speed of one body in relation to the speed of light.

In this regard, a direct relationship between speed and mass has been achieved, leading Einstein to conclude the equivalence of energy and mass. This brought Einstein to formulate his famous equation, where mass is directly convertible to energy and mass is multiplied to the square of the constant existing in the universe, the speed of light.

$$E = mc^2$$

Since the speed of light is phenomenally large, then the energy that could be released will also be enormous. In the same way, if someone wishes to travel at the speed of light, he would need a huge amount of energy but would get heavy as he travels without reaching it because by then he would need more proportionate energy until he reaches the point where he needs infinite amount of energy for he is accumulating infinite amount of mass.

The proposition that nothing can travel near the speed of light, equal or go beyond it is an essential assumption of Einstein's Special Theory of Relativity. Coupled with such a concept is the idea that everything in space is in motion and only the speed of light is constant though its source is moving. Here is where Newton's kinetics breaks down. Since Newton proposes that speed is determined by dividing space over time, without considering the accumulation of mass and dilation of time, then we could divide an infinite amount of space over an infinite amount of time. There would be no limit.

For example, two planets, Centaur and Taurus are separated by a distance of 3,000,000 miles. It is assumed that a rocket can cover the distance in 5 seconds. Since distance is divided by time, then 3,000,000 miles over 5 seconds would be 600,000 miles per second which is the rocket's speed. This is incredible for the spacecraft has surpassed the speed of light.

Moreover, the direct conversion of mass and energy is another departure to Newtonian physics where a body in order to calculate its energy needs to be moved. This is called kinetic energy or energy in motion while a body at rest is not necessarily void of energy but it has potential energy or the energy an object contains to enable it to do work. When such body which is at rest travels, kinetic energy has been used as determined by its mass and velocity.

$$KE = 1/2mv^2$$

But for Einstein, that body which has mass can be directly converted to energy even without moving it. In here comes Einstein's famous equation.

Spectacular, but for the purpose of this discussion more spectacular is what Einstein's Theory of Relativity can do with space and time. Remember that space and time are two separate properties in the Newtonian sense. Space is divided over time to account for speed. Moreover, you are aware of three-dimensions with which you live in. You can move left or right, jump up or down, step forward or backward. You can empirically sense these three-dimensional directions and interpret space and objects in three-dimensional plot. But is there a fourth dimension?

Earlier, the inertial transformation converted the position and direction of motion in a coordinate system but Hendrick Lorentz incorporated relativity in view of time elongation and length contraction (Figure 4.7).

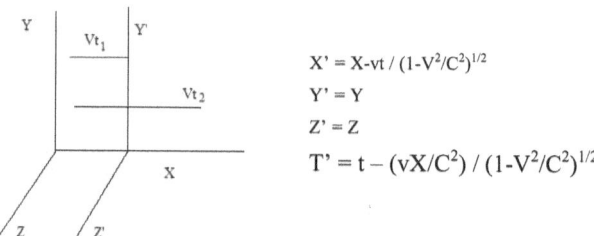

$$X' = X\text{-}vt / (1-V^2/C^2)^{1/2}$$
$$Y' = Y$$
$$Z' = Z$$
$$T' = t - (vX/C^2) / (1-V^2/C^2)^{1/2}$$

Figure 4.7 Lorentz Inertial Frames and Transformation

But though Lorentz considered the phenomenon of length contraction and time elongation in a relative inertial frame, his presumption, just like Newton, is of the absoluteness space and its separation from time. With the relativity of speed, length, mass, and time in relation to the speed of light, Einstein pushed the idea that time is another dimension in relation to space.

You go about your life in three dimensions. But consider Einstein's proposition that everything is in motion. Then the position where you are now will not be the same in the next second. It would not be a problem understanding this if you yourself are moving. But if are just seated then you would think of yourself to be at rest. But the earth is actually rotating. Thus even if you are seated, you are in fact, in motion. Therefore if you move about in three directions (up-down, left-right, forward-backward), you also move about in time. As space is relative in relation to motion, time is also relative in relation to speed. Time then is the fourth dimension.

Since speed is relative and only the speed of light is constant, then we could use such a constant to be the conversion reference of both space and time. Since the speed of light is 186,000 miles per second then one second is equal to 186,000 miles. And one second equals 186,000 light-miles and one mile would take 0.0000053 second for light to cover. One light-mile, therefore, is 0.0000053 seconds and one second is 186,000 light-miles. In view of the Special Theory of Relativity, it seems Einstein waved his magic wand and converted time into a measurement of space. Since it is presumed that time is the fourth dimension, then having translated it into a quantity of space, it could be generalized using the Pythagorean theorem in the likes of the three dimensions done before. But there is another dilemma. Spatial direction is directly progressive. The greater the length covered, the greater the measurement and the greater the number registered. But for time, it is the reverse. The faster you go, the smaller the time registered. The longer the time covered the slower you move. If you generalize this using Pythagorean theorem then, time is negatively added.

$$(-t^2 + x^2 + y^2 + z^2)^{1/2}$$

Again at this point, the separation between space and time has been shattered. Speed is not anymore measured by dividing time from space. But speed is measured by its factor from the speed of light. Distance is not anymore measured in three-dimensional points but in four dimensions with time as another variable, uniting space and time into space-time.

Thus instead of simply determining the distance covered by the bus by multiplying velocity (v) and time (t) in a strictly Newtonian sense, distance and location can be figured out by determining the position of the bus in reference to three spatial points: (Y) the distance of the bus to a reference point on top of it, (X) the distance of the bus to a reference point ahead of it, and (Z) the distance of the bus to a reference point on the side of it subtracted to the time (t) the bus took to cover the distance measured in light distances in the Einsteinian sense. As it cruises along that speed, the bus shrinks in size and gains mass. You become heavy, (the fly as well) and time inside the bus retrogresses all in proportion to the speed of light. As you reach your destination and the bus stops everything goes back, as everything were, when the bus was at rest... with only one difference. You reached your destination but the fly ended its journey a little earlier, as it remained splattered on the windshield, miserably caught by the driver's flyswatter.

Textualizing Biological Life

At this point, we could easily deduce that the physical sciences have all the attributes fit for a milieu-already-constructed. But the living species are also governed by principles which could draw some strokes of a milieu already made. Living species are a complex set of biological system. The human beings inside the bus and the fly splattered on the windshield are all biologically constructed with systems that are complex in operation. Each system, like the digestive, respiratory, nervous, circulatory, or muscular system is made up of specific cells with their unique functions once put together. There is specificity as to the function of each system and its transmission from one generation to the next account for a uniform functioning unit.

The Milieu-already-constructed

Aristotle spent a large part of his research life doing biology but it did not occur to him to ponder in finding out the basic unit of life. This is understandable for he was not a monist who advanced a single entity comprising the many. Aristotle's concern was not the constitution of things but their causes which formed the basis of his metaphysics. On the other hand, the basic unit of life cannot just be contemplated neither simply observed which he methodically did in his taxonomical work in the island of Lesvos for it needed the invention of an instrument to peer through the substance of life.

The first compound microscope was invented by Hans Janssen and Zacharias Janssen in 1595 by positioning two convex lenses in a tube. It was, however, in 1665 when the English inventor and scientist, Robert Hooke who peered through his own assembled microscope the tissue of dead cork and found them to be composed of compartments partitioned by empty spaces. He called these compartments "cells" which he was the first to introduce in the scientific community. The concept was introduced but its proposition as a basic unit has not yet been added.

Since Hooke used dead specimens for his discovery, his colleague Marcello Malpighi studied plant cells and proved the structural presence of cells. Nehemiah Grew, in a separate effort, also established the presence of cells in plants which he described as spaces found in rising dough. In 1676 the Dutch biologist Anton van Leeuwenhoek subjected pond water under the microscope and found microorganisms which he called "animalcules." He continued on with his study and in 1683 published his discoveries by subjecting his own saliva under the microscope and discovered another type of these animalcules which would later on be classified as bacteria. In 1833 Robert Brown discovered the nucleus of the plant cells but it was in 1838 when the German botanist Matthias Jakob

Schleiden generalized that cells are the basic building components of all plants. A year later his fellow German botanist also reached the conclusion that all animals are made up of the basic building blocks called cells. All organisms are made up of cells. And cells are the basic structural and functional unit of all organisms. From here emerged the first two statements of the cell theory. The third statement of the cell theory was supplied by Rudolf Virchow in 1855 when he advanced the dictum *Omnis cellula e cellula* which means "cells develop only from existing cells" (Robinson 2015).

But even if the cell theory was already proposed, the components of the cell would have to await discovery for a compound microscope would not be powerful enough to unveil its constitution. It awaited the development of the electron microscope in order to peer into the physiological components of the cell. The basic unit of life was actually a microscopic functioning system. In it are organelles (little organs) which are also imbued with definite structure and function. If you have skin that encloses your whole body, the cell has cell membrane that acts as protective shield, a means to enclose the cell and its organelles, and a regulatory mechanism to facilitate and filter the entry of nutrients in the cell. Plants and prokaryotic (primitive or simple) cells have another protective covering called the cell wall. Inside the cell membrane is a viscous liquid called cytoplasm, composed of water and complex molecules including sugar, lipids, and amino acids which are responsible for metabolic processes.

Specific organelles with their specific functions metabolize specific substances in the cytoplasm. The endoplasmic reticulum are folds of cell membrane with smaller spherical bodies called ribosomes attached all over its body. These ribosomes are responsible for the synthesis of protein. Throughout the cytoplasm are larger spherical bodies called mitochondria which process carbohydrates to release energy. It is called the power house of the cell. Other organelles are located inside the cell but the most important of which is the nucleus where the chromosomes are contained. How could we picture the cell in relation to its parts?

On your breakfast table, picture your plate as the cell. The edge of the plate is the cell membrane. The pieces of bacon inside the plate which appear like folds of flesh are the endoplasmic reticulum and the sprinkle of salt or pepper are ribosomes. The sausage in your plate is the mitochondria and the sunny-side-up egg is the nucleus. Now having the picture of the cell and some of its parts, which is true to every living thing, what is equally important in the reproduction of life, is that thing contained inside the nucleus... the chromosomes.

The proposition that the chromosomes carry the gene which is the basic unit of heredity is the Chromosome Theory of Inheritance. Walter Sutton and Theodor Boveri first proposed this theory in 1902. The genes therefore are not anymore conceived simply as factors which are expressed in several alternative forms or alleles, but they are rather observable empirical bodies. Thomas Hunt Morgan in 1900 was first skeptical to this theory but he provided one of the convincing evidences to this idea. Instead of experimenting with pea, Morgan worked with fruit fly (*Drosophila melanogaster*), a fast producing insect which is characterized with specific traits. At this time, it was already believed that an organism has certain number (2n) of chromosomes paired (n) with each other. Through microscopic investigation, a human being has 46 chromosomes of 23 pairs. These chromosomes were investigated using the technique of dying a preserved cell at the time of cell division and photographically representing the chromosomes (karyotype) from the largest to the smallest pair. It was also observed that chromosomes empirically exhibit themselves in either X or Y appearance.

Where: A = Red-eyed, a = white-eyed

Red-eyed AA Female	x	White-eyed aa Male
F1: Red-eyed Aa XX	x	Red-eyed Aa XY

F2:   Female

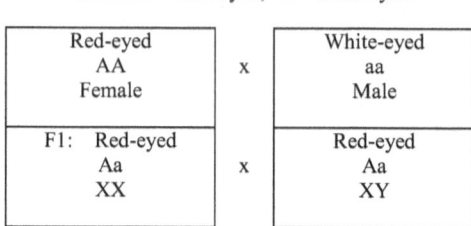

Figure 4.8 Characteristics of the Fly in Morgan's Experiment

It was documented in 1905 that a bug *Protenor's* male gamete has either an X or without an X chromosome, while the bug's female gamete always has an X chromosome. Since different organisms have different number of chromosomes and since an offspring will receive two kinds of gamete from each parent, it was concluded that a female would have a pair of X chromosome (XX) while a male would have only one (XO) at the last pair. The differences in the number of X chromosome would determine the sex of an organism. Later on the Y chromosome was discovered and it was summed up

The Milieu-already-constructed

that a human female would have a pair of X chromosome (XX) while a male would have two different chromosomes (XY).

The Drosophila fruit fly, which Morgan experimented on, has 8 chromosomes. Morgan tried to replicate Gregor Mendel's technique in analyzing hereditary transmission. More of Mendel's theory will be discussed in the next chapter. But unlike garden peas, the fruit flies' sex is characteristically distinct. One would know which fruit fly is male or female. What Morgan did was to mate red-eyed flies believed to be dominant with white-eyed flies believed to be recessive. As the Mendelian genetics proved, the descendants were red-eyed. Later, Morgan varied the experiment. He separated the red-eyed males from their red-eyed sisters and mated them. The results were white-eyed males and red-eyed females (Figure 4.8)[1].

It would appear then that the red-eyed allele was present in the X chromosome with which the female had a pair. This strengthened the belief that the chromosome carries the gene which exhibit the phenotype or the physical characteristics of that gene.

Chromosomes (colored bodies) first appear like very fine threads when the cell is not undergoing cell division. These threads are called chromatids which are first believed to be made up of molecules of proteins and nucleic acid. During cell division, the chromatids begin to coil and condense until they thicken and shorten. It is like having a fine piece of thread and rubbing it between your palms until it thickens. As the chromosome tightens, it constricts at a certain portion called the centromere. It divides the chromosome into two arms. Picture two threads now which you have tightened to

---

[1] To facilitate analysis, Punnett square (after Reginald Punnett) is used.

form a thick spindle and tie them together somewhere along its length. The knot then serves as the centromere. The centromere is the location where the chromosomes split during cell division so that each divided cell will receive equal number of chromosomes. The location of the centromere also defines the kind of chromosome besides its size. If the centromere is located near the center, it is metacentric, if toward the end of one length of the arm, it is acrocentric, if the centromere is near the end, it is telocentric. Some chromosomes have secondary constrictions at the end which form chromosomal satellite. Now a chromosome when karyotyped and stained form bands like the barcodes you see stamped on a product sold in the supermarket. Each band represents a certain gene. The differences in the bands determine the differences in the chromosome.

Now besides the arm and centromere, a chromosome which has been karyotyped, can be analyzed according to its parts in order to locate the genotype or the allelic constitution in the gene that gives rise to the physical trait or phenotype observable in the organism. The upper arm (above the centromere) of the chromosome is designated as p while the lower arm is designated as q. Each arm is separated into regions in numeric form ascending or descending from the centromere. Each region is further separated into numbers which designate each band.

Thus if you look at the two chromosomes (Figure 4.9), you will find that at locus or location p6 the two chromosomes have a different allelic constitution or genotype (Aa). At q2, the two chromosomes are again made up of different alleles (Bb). In these two locations, the two chromosomes are said to be heterozygous, but at locus q4 the same alleles are situated (cc). In this location, the two chromosomes are homozygous. It could then be concluded that the genotype represented by these chromosomes would be AaBbcc which could represent observable physical traits or phenotype.

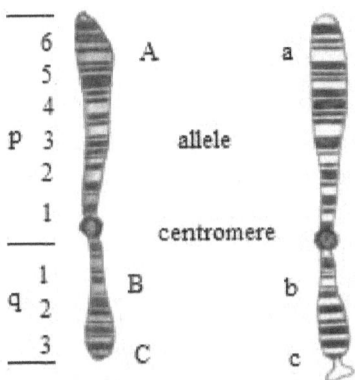

Figure 4.9 Representation of Two Chromosomes

Now that the morphological structure of the chromosome is known and the idea that it carries the particles that correspond to the expression of physical traits is established, the geneticists next move was to locate the specific patterns in the chromosome and its correspondence to specific character traits of certain organisms. But the question remains, what is the chromosome made up of? This would mean figuring out the nature and structure of the genetic material itself. Remember that the chromatids are like fine threads before they coil and they thicken until they assume certain chromosomal structure.

At this point, scientists are already venturing into molecular genetics. We all know that the basic unit of matter is the atom. Atoms combined together will form molecules. The combination of the molecules of certain elements would produce compounds. Water (H_2O) has two molecules of hydrogen and one molecule of oxygen. It was also discovered, as early as Mendel's time, that organic materials are rich in carbon molecules. These carbon molecules are abundantly found in living organisms. Since chromatids were hinted to be made up of molecules bonded together like fine threads, the genetic material would then be nothing but molecules of certain organic compounds.

As early as 1869 in Germany, Friedrich Miescher was able to isolate nuclei from pus cells attached on waste surgical bandages. He found out that the nucleic substance contained phosphorus called nuclein. In 1928, the foundation for the unraveling of the genetic material was laid. Frederich Griffith performed his transformation experiment on the pneumococcus bacteria known as *Streptococcus pneumoniae,* responsible for the spread of pneumonia. There are two strains of this bacterium. One type has smooth coating which, when viewed in a microscope, appears to be a spherical cell encapsulated with mucus coat that acts as its insulator. This cell is lethal for its mucus protects it from the attack of antibodies. The other strain is rough when viewed in the microscope for it lacks the capsule that defends it from the attack of white blood cells. What Griffith did was to inject the smooth coccus into a mouse. The mouse died of pneumonia and when dissected, the dead mouse bore the smooth strain. Another mouse was injected with rough colonies and the mouse survived bearing no trace of the bacteria after it was dissected. This is proof that the mouse's white blood cells destroyed the bacteria. Next, he subjected the smooth strain to heat and injected the colony to the mouse. The mouse lived and no trace of the bacteria was found in the mouse. This proved that heat dissolved the mucus coat which made the bacteria vulnerable. Lastly, Griffith injected the mouse with both the rough strain and smooth strain which was formerly subjected to heat. The mouse died of pneumonia. And when dissected, the mouse only bore the rough strain but this time bearing the protective capsule. Griffith's conclusion was that the rough strain (non-lethal) was transformed into smooth strain (lethal) via the inheritance of the mucus capsule which the smooth strain that was subjected to heat formerly bore. The heat destroyed the mucus but the gene which was responsible for the production of the mucus capsule was passed on to the rough strain and made it lethal. The next step then, would be to unravel the chemical composition of this substance responsible for carrying the inheritable trait. The discovery of the chemical composition would reveal the nature of the genetic material which is the major component of life (Weaver and Hedrick 1991).

One of the missing pieces of the puzzle, however, was supplied by Oswold Avery, Colin Macleod, and Maclyn McCarthy in 1944. Their objective was to define the transforming substance in virulent cells which Griffith characterized. Using Griffith's transformation experiment as basis, they first destroyed the protein present, but the virulent cell still retained its transforming quality. This ruled out protein to be the transforming substance. Then they tried deoxyribonuclease (DNase) and the transforming effect ceased. The experiment gave one conclusion, DNA or deoxyribonucleic acid is the missing piece of the puzzle. It is the genetic material being sought for. Further tests proved the findings. The transforming substance was subjected to ultracentrifugation by spinning the substance in a centrifuge. The substance rapidly sedimented at the bottom of the centrifuge. This suggested that the substance was of high molecular weight characteristic of DNA. They also placed the substance in an electric field (electrophoresis) to see how rapidly the substance moved. The substance was found to be highly mobile just like DNA. Next, it was placed under spechtrophotometer (ultraviolet absorption spectrophotometry) to determine the kind of ultraviolet light it absorbed. It strongly absorbed the light with the wavelength of 260 nanometers. Later, elementary chemical analysis yielded an average of nitrogen-phosphorus ratio of 1.67 (Weaver and

Hedrick 1991:134). The assumption that the genetic carrying molecule of inheritance is a chemical called deoxyribonucleic acid is the DNA theory.

By the end of 1940 the chemical structure of DNA had been unlocked together with RNA, ribonucleic acid, a genetic material present in viruses. The DNA was found to be primarily composed of substances which can be grouped into three groups: nitrogenous bases, phosphoric acid and sugar in the form of deoxyribose. The nitrogenous bases are of four types: adenine (A), cytosine (C), guanine (G), and thymine (T). The nitrogenous bases are, therefore, called the alphabet of life. They are categorized into two types of compounds. One is purine which includes adenine and guanine and the other is pyrimidine which consists of thymine and cytosine. The RNA, on the other hand, contains the same phosphoric acid but it has ribose as its sugar content rather than deoxyribose. Moreover, one of its nitrogenous base thymine is replaced by uracil (U) under the pyrimidine group. But even if the alphabet of life has already been discovered, the question remains as to how the molecules of these compounds are bonded together and arranged? The race towards the structure of DNA was on.

Linus Pauling first showed interest on how the DNA molecules look like. A theoretical chemist that he was, his contribution was the helical structure of protein molecule. By helix, it appears like a thin strip of paper you wrap around your finger. Slide the piece of paper and it still coils like a screw. Another piece of information was provided by Erwin Chargaff when he proposed that the content of purine equals that of pyrimidines. The amount of adenine equals the thymine as the amount of guanine equals that of cytosine.

But the crucial piece of the puzzle was supplied by the tandem of Maurice Wilkin and Rosalind Franklin and their colleagues. They used X-ray diffraction to uncover the three-

The Milieu-already-constructed

dimensional structure of DNA. The technique was employed by preparing a viscous solution of DNA and pulling a fiber containing a batch of DNA. The fiber appeared like diffracted crystal using X-ray. But story has it that Wilkin had an uneasy relationship with Franklin. Wilkin showed the X-ray photograph over dinner with James Watson and Francis Crick who was theorizing on the structure of the DNA using other scientists' findings. The X-ray picture showed a simple image of dots that condense into circles at the outside but forms an X toward the center. In contrast, the X-ray diffraction of protein molecules appears as a set of dots like a shotgun blast (Weaver and Hedrick 1991: 140). The X-ray picture of a DNA, therefore, would be a simple structure that repeats itself like a spiral. It would appear like stretched coil of spring which when viewed at one end would form concentric circles. This confirms that the spiral appearance in the X-ray crystallographic image is a helix. But there is one point to resolve. How could the Chargaff's rule be applied? Chargaff's rule states that nucleic base purines equals pyrimidines. If there is one helical strand, then one purine at one end will not have a counterpart on the other.

Watson and Crick found a way to solve the problem. Instead of one helix, two helices serve as the backbone of the macromolecule strand with the nitrogenous bases in between. Picture a ladder where two long parallel poles are connected by smaller lateral steps. Then imagine twisting the ladder. The two long poles at each end create the double helix and the small lateral steps represent the purines and pyrimidines in equal amount (Figure 4.10).

But there is still one issue to resolve. How does this complex macromolecule called gene able to carry the hereditary information needed to pass it on to the next generation? It should be clear by now that the gene should contain all the information needed to direct the growth and life of an organism. Like a book or volumes of books, the gene with its code is the repository of how an organism maintains life and how it develops through time. In order to direct life and its growth, the DNA code should be spread on all parts of the organism's body. The code is then translated into usable form on all tissues of the body. Moreover, the DNA should be capable of replication and a copy of which is transmitted to its descendants.

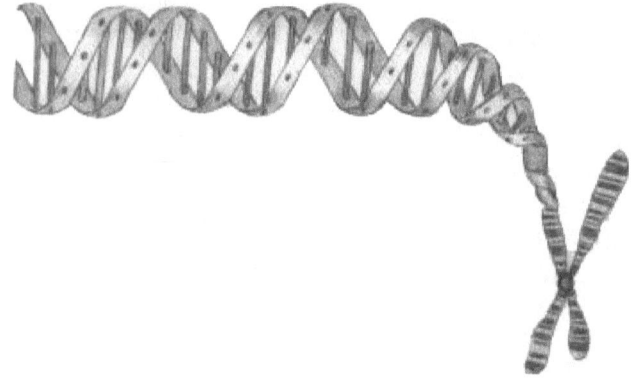

Figure 4.10 DNA Spindled in a Synapsed Chromosome

Even before the discovery of the DNA, the body has been found to compose large amounts of protein. Different categories of proteins have definite functions in the body. Enzymes are proteins which act as reaction enhancers or catalysts. Some proteins are structural or building blocks of body tissues. Keratin, for example, is a protein present in the skin and hair. Muscle tissues are built with proteins. The antibodies or white blood cells which protect the body from harmful elements and the hemoglobin or red blood cells which carry gasses in the blood are special types of proteins. Hormones that provide communication to the organs of the body are proteins as well (Maxson and Daugherty 1989: 159).

Proteins are also complex macromolecules consisting of amino acids. Twenty types of amino acids have been found in most species. Twelve of these are synthesized by the human body and the remaining eight are supplied by everyday food intake. Protein molecules have two amino acids bound by a peptide bond. The differences depend on the chains of polypeptide bond and its combination with other molecules. Hemoglobin, for example, contains four polypeptide chains attached to an iron-containing molecule (Maxson and Daugherty 1989: 159).

If DNA has the genetic information and protein is the building block of organisms, how does the genetic material create the message and how does it relay it to protein molecules so proper tissues could be constructed? To answer this, we have to answer first what is the genetic code? So far, as you have reached this part of this project, you may have recognized the individual letters and figured out the meanings of letters constructed into words and words combined into sentences. In order to understand the meaning of a word, we look it up in the dictionary. But the word itself is only a combination of letters. If a word is a combination of letters with meaning, the genetic code is a combination of nucleic base that create certain messages. Remember that DNA helices have two deoxyribose and phosphate molecules on each side as backbone and nitrogen base of two categories, purines and pyrimidines. Purines have two molecular types adenine (A), and guanine (G) while pyrimidines are also of two categories: thymine (T) and cytosine (C). This is the alphabet of the gene or the alphabet of life whose combinations between purines and pyrimidines produce certain messages. The next question is, how does the combination of the nitrogenous bases produce coded messages?

If we unwind the coil of DNA strand just like the a thin thread, two long parallel molecules of deoxyribose and phosphate molecules produce a backbone of four nitrogenous bases adenine (A) and guanine (G) which are purine bases and thymine (T) and cytosine (C) which are pyrimidine bases. Since the DNA alphabet consists only of four letters, it would not be enough to create 20 different messages and instruct 20 amino acids to produce proteins if a code would only consist of a double pair ($4^2=16$). But a triple pair combination ($4^3=64$) would be sufficient. This triple pair is called codon or a word in our language (Maxson and Daugherty 1989: 159).

But what does this codon or triple pair combination mean? This would mean decoding a whole range of triplets or finding the meaning of every word in a dictionary. To answer this, the discovery of another type of ribonucleic acid (RNA) was in place. It would be recalled that the DNA is situated in the nucleus of a cell while ribosomes which are attached to the endoplasmic reticulum that synthesize protein are located in the cytoplasm outside the nucleus. There ought to be another medium which the code contained in the DNA could reach the ribosomes. Here the messenger RNA (mRNA) comes in. Unlike the DNA, the ribonucleic acid (RNA) is a single strand molecule with ribose as its sugar content rather than deoxyribose. It has the same adenine, guanine, and cytosine as nucleic base but instead of thymine, uracil (U) is the other pyrimidine base. It being single helix rather than double strand has its own purpose. What the messenger RNA does is to transcribe the code found in the DNA with its own nucleic base. The transcription is then delivered to the ribosomes in the cytoplasm for the production of proteins with the assistance of transfer RNA (tRNA) (Maxson and Daugherty 1989, 171). Therefore, the genetic code of three base pair would only be meaningful if the protein it produces with the RNA transcription is also discovered.

Thus working on the idea that the dictionary of the gene consists of a triple two-paired nucleic molecules, M.W. Nirenberg in 1960, produced a mixture of 20 amino acids of

Escherichia coli in a test tube. Then he synthesized an artificial message consisting of uracils (UUU). He examined the polypeptide produced and found only one type of amino acid, phenylalanines. Thus the code UUU instructs the production of phenylalanine amino acid. Similar experiments were done and he found out that AAA encodes lycine, CCC instructs the production of proline and GGG sends the code for the manufacture of glycine. If the English language begins a sentence with a capital letter and ends in a punctuation mark, the codon AUG creates the amino acid methionine which is known to be the initiator codon for messages that instructs the production of protein. It is the capital letter in every sentence. On the other hand, the codon UAA, UAG, UGA were found to terminate the message. Thus they serve as punctuation marks (Maxson and Daugherty 1989, 171). The right code therefore would produce the right protein but some aberrations in the code may not create the proper polypeptide needed by the organism, which in the process results in certain types of diseases. It was also discovered that the genetic code has a universal feature. The UUU code produces the same phenylalanine in *E. coli* bacteria, humans and other organisms.

The DNA, therefore, is the universal component or kernel of life. All living things are constructed with it. The genetic code is life textualized. This is then an aspect of the milieu-already-constructed that composes biological life.

Textualizing Economic Relations

Suppose you visit a store to look for the best onions for your Mexican dinner. In a certain stall you're confronted with different kinds of onions but noticeable among them are the two types, the big white onions almost the size of billiard balls and the little red ones the size of ping pong balls. You have already known that the differences in these onions have been the result of the differences in the DNA makeup that account for the dissimilarity in their genetic characteristics. But your concern is not to ask for the chromosomal differences nor ask the saleslady for the genetic pattern of the two onions. Your concerns are the prices and quantity of the onions you will take home. The big white ones, however, are not as spicy as the red little ones. You will use the bland, big white ones for the salad and the spicy, red little ones for sautéing. The prices of these onions, however, vary.

In simple economic analysis, what we have is a market. You are a buyer in need of a commodity which you could secure from a seller. Thus you find three components of a market, which are quantifiable in three variables: the buyer which dictates the demand, the seller which provides the supply and the commodity which is secured through its price. Two weeks ago the price of one kilo of the white onions was twice as much as the red ones. Now the price of the red ones is 0.5 times higher than the white ones. Now you may ask, how come prices vary?

First, we have to suffice ourselves with some prescriptions on how the market works. Economics is concerned with scarce resources. Onions may not be available if farmers prefer to plant other more profitable crops. Even the possibility of having scarcity of people who will prefer to work on the soil and produce crops is not remote. Thus if resources and labor are scarce, the means to allocate them would be through the value they can be secured. And this value is through the price as represented by a negotiable instrument called money. In a working market therefore, the demand, supply and price are quantified through the amount of money they can be exchanged. But even if we are dealing here with resources, commodities and money, economics is in fact concerned with human relations and human behavior. Economics is the relation between buyers and sellers at instances during which buyers buy and sellers sell certain commodities at a certain price through a negotiable instrument. Thus human relation is quantified through money, the power of an individual to participate in human economic relations is valued through money and all these can be secured through the price in which commodities and services are available.

Second, the management of scarce resources leads to only one condition and that is "full employment" of all the factors of economic activity. This would mean that all the factors production and consumption are utilized to their utmost that wastage is minimized.

Third, economics assumes that human beings who are engaging in this relation are rational. This means that an individual would maximize his gain and minimize the cost in securing his desired commodity or service through his limited resource. He would procure the best commodities at the lowest price possible and even produce certain goods at the lowest possible cost. He would try to lessen his losses and make sure he would generate enough or more profit by producing certain commodities or rendering certain services. Scarce resources then are to be managed in production and use. But who would manage these scarce resources except by individuals who are rational. Economics then is concerned with rational individuals managing limited resources that they produce and use.

Fourth, there is an underlying and scheming concept of what economics is about aside from the management of scarce resources by rational people. This is the allocation of wealth. The irony is that rational people who are constrained with managing exhaustible resources are presumed to be concerned with acquiring and accumulating wealth. They try to accumulate wealth out of the resources which are limited. It is this third assumption that the development of the different schools of thought in economics emerged.

Land may have probably been the oldest means to produce surplus out of something that human beings need or want. Agriculture is the closest means to guarantee the survival of human beings for food is the very stuff that would ensure the perpetuation of life. But no one can plant or farm without land. Land, therefore, is a source of wealth. The bigger the land, the larger the wealth. But as the farm products that the land produce are exhaustible, productive land that would supply the farm products are also scarce. The competition for wealth through the ownership of large tracts of productive land would end up to one which possess power by charisma, influence or brute force.

The symbiosis of power and land fuels the feudal system. And if the most influential, charismatic or brutishly powerful accumulates more land and proclaims himself king as connived by the church to confer to him the kingship, then the whole kingdom becomes a huge serfdom, where the king is the ultimate landlord and his subjects are his serfs. The wealth of the kingdom will definitely be the wealth of the king. The bigger the territory, the bigger the wealth and the richer would be the king. It was incumbent for the king to expand his territory even at the expense of invading other lands simply to accumulate wealth. This is the heart of the feudal system which was the economic system at work during the medieval age. But one city which defied this assumption was Venice in late 14th century.

Venice, which could be considered a state among other states in the Italian peninsula during the Renaissance, was permitted by the Ottoman Turks to trade with the Asian territories in the Mediterranean Sea. Venice acquired riches, not by producing surplus from its land, but by buying from other territories and selling them in other places. This emerging economic system suggests that the kingdom should act not like a farmer but behave like that of a merchant. It assumes that the kingdom can be wealthy and the king with it, if it buys and sells products which it may not necessarily produce. This is mercantilism.

With the entry of Portugal and Spain in finding sources of commodities most especially spices overseas in the 15th century, the desire of kings to acquire wealth came with monopolizing the market and sources of goods that would sell high in the European market. And in order to monopolize the market, they colonized these territories which shut the market of these commodities to other nations but exclusively opened it only to European colonizers that occupied them. Mercantilism, therefore, as practiced during the Age of Expansion in Europe would presume that not only was the boundary of its kingdom be the land of the king but the land they colonized overseas in Asia or America be also owned by the king. Mercantilism then became feudalism overseas. It was highly a government or king-instituted and dictated economic system. The market overseas was a king's possession and no one can just import or export from the colony without the king's permission. Trade, in the spirit of liberty, would not just be permitted by the king since merchants trading with other nations might leak the wealth which was supposed to be siphoned to the king. Trade would have to be one-way – more going in for the kingdom, that is more that it would sell or export overseas and less that it would buy or import from others. Thomas Mun expressed this position in his book *England's Treasure by Forraign Trade* published in 1630.

> The ordinary means therefore to increase our wealth and treasure is by Forraign Trade, wherein we must ever observe this rule, to sell more to strangers yearly than we consume of theirs in value (Sandelin, Trautwein and Wundrak 2008: 10).

This state policy proved defeating for merchants which clamored for freedom to do their business at home and overseas. This very idea came under attack by a liberal in the name of Adam Smith who published his book in 1776, *Inquiry into the Nature and Causes of the Wealth of Nations*. Thus was born the Classical Economics which espoused free market, with the assumption that freedom as infused in the market where buying and selling transpire is the best way to allocate resources. The "invisible hand" of the market, a system, where prices of commodities with reference to the quantity of goods being transacted is arrived at without the government intervening to set the price or impose the volume of goods being sold.

With price as a means to allocate resources, it would mean we have to put monetary value on our efforts so we could have the means to secure the commodities we need. This being the case, then, how did the onions get its price? Let's begin with the farmer. A farmer cannot plant onions unless he has the space with the proper soil for the crop to grow. Of course, the farmer should be equipped with the proper technology, seeds, fertilizers and other necessities required to prepare the soil, plant the seeds, nurture them, until they get harvested. All these factors will add up to the cost of his onion production including the rent or monthly amortization for the land he is cultivating. All these are monetarily quantifiable. The cost of rent, seeds, fertilizer, even hired labor are all measurable even to the last cent. And these costs will add up to the price of the onions once he sells them in the market. But the question is, how much would the farmer add to the cost of producing onions commensurate to his own labor. How could he quantify his efforts, his sweat, and transform them into monetary value which could mark up the price of the onions?

In this scenario, Adam Smith assumed that the value of a commodity is derived from the labor spent in producing it. Labor, according to him, is the real measure of the exchangeable value of all commodities. Labor, however, is ascertained by virtue of the time spent, hardship and ingenuity in producing the commodity. The greater the time spent, the harder it is in producing the product and the more intricate the need to produce it, the higher the value for labor attributed to the manufacturing of the product. With labor as the real measure of price, the quantity of money, on the other hand, is the nominal price (Smith 1776: 13-14).

> In this popular sense, therefore, labor, like commodity, may be said to have a real and nominal price. Its real price may be said to consist in the quantity of necessities and conveniences of life which are given for it; its nominal price, in the quantity of money (Smith 1776: 14).

Knowing that the value of a commodity can be gauged through the labor spent in producing it and other costs, there is still a difficulty in measuring how much the price of labor can be quantified. In this regard, David Ricardo added another dimension in this scheme by assuming that labor has both a natural and a market price. The natural price of labor depends on the price of food, necessities, and conveniences required to sustain the life of the laborer. The market price of labor, on the other hand, depends on the supply and demand for it. Thus if a farmer needs to hire someone who would plow his field and there are many farm helpers who could provide the needed labor, then the tendency for the price of hired help would go down. Ricardo then asserts:

> It is when the market price of labor exceeds its natural price that the condition of the laborer is flourishing and happy, that he has in his power to command a greater proportion of the necessaries and enjoyments of life… When the market price of labor is below its natural price, the condition of the laborer is most wretched. Then poverty deprives

him of those comforts which custom renders absolute necessaries (Ricardo 1817).

Thus if hired help on the field competes for very few work, the price of labor goes down but if there were a lot of work in the field with very few workers for hire, then the price of hired help could go up. If the wage of hired help, which is the market price of labor, is higher than the accumulated price of his necessities, which pertains to the natural price of labor, then he is well off, but if he is paid less than the aggregate amount he could buy for his necessities, then he could be in for a lean day.

Thus for a farmer, he would have calculated his efforts through the amount of money he needs to sustain his everyday necessities. He could even add the cost of his production and even add more if he desires extra profit. But all the aggregate prices which the farmer included in the sale of the onions would all depend on how the market would accept the price the farmer offers. As the farmer places his onions in the market, the Law of Supply and Demand gets to work.

Smith had long contended that if the market absorbs the price the farmer wishes for his produce considering the labor and cost of production, then the commodity gets its natural price. But hardly does the market accommodate the natural price the farmer wishes for his produce to be sold, for the market still sets its own price.

> The market price of every particular commodity is regulated by the proportion between the quantity which is actually brought to market, and the demand of those who are willing to pay the natural price of the commodity, or the whole value of the rent, labor, and profit...(Ricardo 1817)

At this point, the Law of Supply and Demand is at work. The price of onions would be determined by the quantity of two variables: the quantity of the demand or the amount of those who need the onions, including you, and the quantity of the supply or the amount of the onions available which the farmer delivered to the market.

The Law of Supply and Demand states that the price of commodity soars if there is too high a demand for it but there is not enough supply to sustain the demand. The price of the commodity plunges if there is an oversupply of a commodity without enough demand to buy it. The price of a commodity, therefore, is inversely proportional to the demand, holding other factors constant. As the price of a commodity increases, the buyers become reluctant to buy it. But as the prices of goods and services slide, the buyers would be more eager to secure the commodities and services for they would be at an advantage to secure them a price they could afford and even garner enough savings.

Adam Smith, David Ricardo, Thomas Malthus, Jean-Baptiste Say and John Stuart Mill with their individual works can be grouped together in the classical school of economics. This economic school is more philosophical, establishing the principles of economics rather than presenting the discipline in a mathematical approach. Another school of thought that placed these economic principles in a mathematically practical schematics was the neo-classical or the marginalist school. This group of economists assume that the rational person behave in margins in pursuit of economic maximization. It was Alfred Marshall with the publication of his book *Principles of Economics* in 1890 that graphically illustrated the dynamics of price and quantity with the opposing interest of consumers and suppliers.

The famous representation of the oppositional behavior of buyers and sellers is in the famous graph (Figure 4.11).

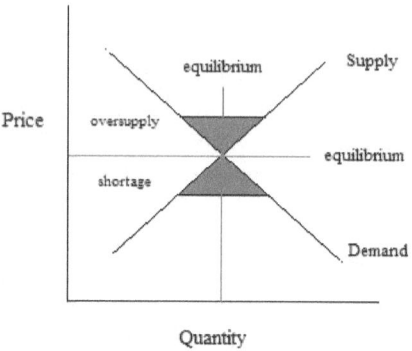

Figure 4.11 Marshall's Classic Supply and Demand Curve

The price of a commodity is directly proportional to supply considering other factors constant. This means that as the price goes up, the producer would likely deliver more of his produce to the market to take advantage of the high price but the buyer would always be hesitant to buy if the price of the commodity is high. The buyers and sellers, therefore, being two rational individuals would have two contradicting interests which would permit their interests to meet at a certain level. This is called the equilibrium where two lines in an X and Y plane intersect each other.

> When the demand price is equal to the supply price, the amount produced has no tendency either to be increased or to be diminished; it is in equilibrium. When demand and supply are in equilibrium, the amount of the commodity which is being produced in a unit of time may be called the equilibrium-amount, and the amount at which it is being sold may be called the equilibrium-price (Marshall 180: 345).

If the price were set higher than the equilibrium, an oversupply results for only few buyers would hazard a high price of the commodity (Figure 3.10). If the price, on the other hand, were set below the equilibrium price then a shortage of supply results for there would be a sudden increase in buyers eager to purchase the cheaper commodity (Mankiw 1991: 76-77).

The market then and a free market at that is the best way to allocate scarce resource or manage it. The invisible hand of the market of the classical school and the supply and demand curve of the neo-classical school are a means to analytically illustrate the operation of the market. Short supply would naturally catapult prices. This is natural for if prices are kept low with supply shortage, then consumers would buy more what they need for fear that they would not have enough because of the short supply. A disastrous outcome would ensue. Supply would be exhausted come the next day and there would be nothing left for others to have what others need to be supplied with. If prices go up of a commodity in short supply, then it would limit what individuals could afford to buy, therefore apportioning the commodity in short supply to as many people who could afford them at the moment. A supply in abundance, however, depresses the price in order for more people to buy it, therefore exhausting the produce. Here is "full employment" of the factors that produced the commodity and the commodity itself on sale is achieved.

When typhoons devastate a place, prices of goods in that region go up. This is natural and should not be feared because that is the natural way of the invisible hand of the market at work to manage the scarce resources available. Human beings, rational as they are, will naturally hoard the commodities they will survive with. If prices are kept low, people will buy more than what they need, depriving others of the supply. Retailers will hoard with prices that are going up for if they would not do us, then consumers would flood on their stores and would buy up all the supplies depriving them of what to sell the next day and placing them out of business.

The government, if they have the supply, will also hoard and ration the commodities in short supply for it would unnecessarily expend all the goods in short supply at their disposal.

The demand curve assumes the inverse relationship between the price and the quantity of goods bought: that with less demand, the market is flooded with goods and the prices go down but with high demand there would be lesser quantity of goods and prices go up. But there lies beyond the demand curve and that is the satisfaction derived from the goods they demand. With this concern, Marshall developed his Marginal Utility Theory which assumes that satisfaction (he termed utility) can be quantified in the same manner as measuring physical objects. Marshall assumes that a person who consumes a certain commodity incurs satisfaction where the sum of individual utility or satisfaction over time is called total utility and the added satisfaction derived from the addition of one product is called marginal utility.

Marshall further assumes that as an individual increases his consumption of a certain product, he may reach the peak of his satisfaction and with the addition of the same product the less satisfied he becomes. It is like eating ice cream. An individual may initially consume pint after pint of ice cream but he may get used to it that his desire to eat more ice cream would be less in the succeeding moments of consuming it. This is Marshall's Law of Diminishing Marginal Utility.

Marshall further assumes that besides being rational, an individual does not have unlimited income that he would necessarily try to maximize his money to buy the merchandise he is satisfied with. In the same manner, he assumes that with the consumer not having infinite satisfaction with the same merchandise over time, he could switch to another product that would satisfy his preference. Thus the entire occasion can be considered in equilibrium when it is not possible to switch the least amount of money from product X to Y and obtain an increase in total utility considering the individual's income and the price that the consumer faces. (Ison 1992:48). The condition can be translated into an equation in the following manner:

$$MU_x / P_x = MU_y / P_y$$

This equation means that the Marginal Utility (MU_x) obtained from product X over the Price (P_x) it was bought equals the Marginal Utility (MU_y) achieved from product Y over the Price (P_y) it was purchased, considering that with these two products, a consumer will not reallocate his resources to increase his total satisfaction, since the satisfaction obtained from the two products have already been obtained.

But this model of satisfaction has its own problems. Satisfaction measured in utils cannot be measured in the same manner as length or quantified in the same way as any measurable quantity. Satisfaction is very subjective that there is no instrument to standardize it. We don't have any weighing scale to measure satisfaction. One's satisfaction is personally regarded by himself and he can set his own standard for it. The way a person would set his utils may be different from any other consumer and the distance between 2 utils would vary from one person to another. Utility or satisfaction would, therefore, be different entirely from the standard meter stick or weighing scale where the separation of each unit is standard for all scales.

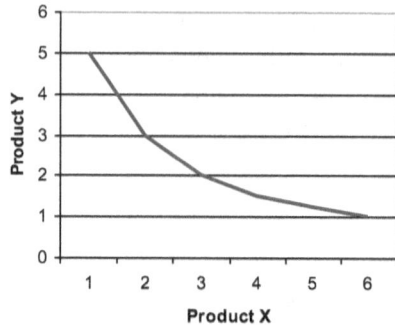

Figure 4.12 Indifference Curve with the Combination of Two Products

This led John Hicks to develop his Indifference Curve Analysis. Since satisfaction is hardly measurable, Hicks argues that preferences could be ordered or ranked rather than measured. Let's say a consumer prefers this number of product X to this number of product Y. This means that satisfaction is determined in comparison to two or more products. Hicks assumes that a convex curve will be produced if the quantities of products X and Y are plotted against each other (Figure 4.12).

The indifference curve, however, reveals the order of a consumer's satisfaction but it does not tell us which combination a consumer would likely choose, for in all combinations, he receives certain level of satisfaction. Hicks adds another component which would likely predict a consumer's decision and that is his income and the price of the product (Figure 4.13).

The Milieu-already-constructed

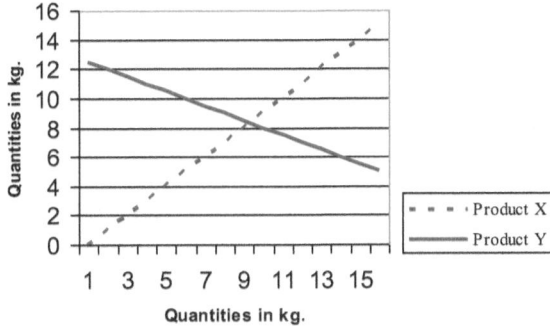

Figure 4.13 Combination of Products X and Y Plotted against the Indifference Curve

Where the two lines intersect, Hicks theorizes this to be the best condition where a consumer could maximize his money to buy a combination of the two products and achieve the best level of satisfaction as constrained by how much his budget could offer. Here is equilibrium attained.

Several factors may affect these lines however, which will shift the lines left or right. An increase in income, for example, will permit the drawing of another budget line to the right parallel to it. Equilibrium will also shift generating different proportions of products X and Y while still maintaining the level of satisfaction. The price of the commodity may fall, letting the budget line to pivot and not shift. But with all these factors, we have to bear in mind that despite the attempts to model satisfaction in X and Y axes, satisfaction is still subjective. Hicks' Indifference Curve may assume an arc different from the arc that would represent someone else's level of satisfaction. Likewise, an individual may change his preference over time. An individual, rational that he is, who maximizes his resources to achieve the highest level of satisfaction, could re-invent himself according to the constraints of his resources. Even if he is at the best position to achieve the equilibrium, he may still opt to move a little below or above the level which the equilibrium prescribes. The same would be true to a producer who would re-invent himself based on the constraints of his capital to manufacture the most profitable merchandise or render the best service his capital could generate.

If the Theory of Marginal Utility and the Indifference Curve illustrate the behavior of consumers on the demand side, the behavior of the firm which the Theory of Perfect Competition is concerned with the supply side of the demand-supply graph. Perfect competition in a free market is assumed to take place when there a) a number of sellers in the market offering b) homogenous products which are perfect substitutes of each other, c) which would lead the sellers to be price-takers for no single seller could influence the price, and that d) there is freedom to enter and exit the market. No single firm could dictate the price of the commodities which are the same in promise for buyers can just switch to another product that would offer a lower price. Since buyers can just prefer the cheaper commodity, the increasing demand would hardly be fulfilled since there would not be enough to supply the high demand for the cheaper product. Those who could not get hold of the cheap commodity would revert back to buying the product sold a little higher.

Operating within the same assumptions, the Theory of Perfect Competition is a theory of production illustrating how a firm would behave in view of maximizing the acquisition of profit with respect to the price of the commodity that it would offer. The price is crucial for in a free market, buyers are sensitive of the price of the commodity they are buying and the price of the commodity would be the reference of how much profit the firm could acquire. The problem for a firm would be: at what price would it offer its product, at what quantity would it produce the product in order to maximize the

The Milieu-already-constructed

gain in profit and if competition worsens, at what price could it be lowered while still taking profit and at what level would the loss of profit occurs thereby leaving the market?

The Theory of Perfect Competition has only one rule to answer the firm's concern. "Profit is maximized at the price (P) level where marginal cost (MC) equals average revenue (AR)."

$$P = AR = MC$$

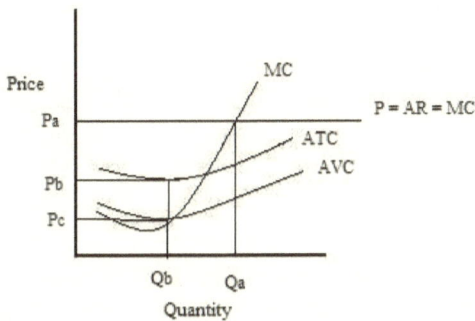

Figure 4.14 Graph of a Firm Operating in Perfect Competition

Marginal Cost (MC) (Figure 4.14) is the differential increment of the amount of expenditure the firm spends in the production of a single unit of product, Average Revenue (AR) is the gain derived for a single product, Average Total Cost (ATC) is the sum of the variable and fixed expenditures that the firm spends per unit product and Average Variable Cost (AVC) is the varying amount of expenditure that the firm spends for a unit of product produced. Since perfect competition assumes that there is freedom to enter and exit the market, the firm begins competing in the market when the price of the homogenous commodity it would sell falls above the region of the ATC. The firm begins losing profit when the prices fall between the regions ATC and AVC that is between

Pb and Pc. Within this region, the firm will scale down operation in producing at only Qb. The firm totally exits the market when prices fall below AVC or Pc which will spell bankruptcy for the firm. On the other hand, profit is maximized at Pa with the volume it would produce the product at Qa. This is the equilibrium region where full employment is achieved.

> Because in any competitive equilibrium with production, the competitive price is equal to the short-run marginal cost of each firm (and in any long-run equilibrium, price is equal to long-run marginal cost), no further gains from trade are possible. No customer is willing to pay what it costs any firm to produce an additional unit of output. This result is a reflection of the *Pareto optimality* or *efficiency of competitive equilibrium* (Eaton, Eaton and Allen 2010).

In reality, perfect competition is hardly achievable. Two books were published in 1933 that would account for the imperfection of the market, *The Theory of Monopolistic Competition* by Edward Chamberlin and *Economics of Imperfect Competition* by Joan Robinson. Variants to a market characterized with perfect competition came as a result only a few sellers producing the identical products or many sellers offering slightly different products. Thus oligopolistic and monopolistic competition became the result. But while the Theory of Marginal Utility could account for the satisfaction of consumers in their choice of products and while the Theory of Perfect Competition could account for the level of production and at what price the firm would sell its product, is there a theory that could account for the competition as buyers buy and sellers their produce in the market? The answer is the Theory of Time, Information and Money.

> The theory of time, information and money is a theory of competitive market which assumes that a

market is marked with competition if similar or slightly dissimilar yet competing commodities ($X_{1...n}$) in the market have more or less equal probabilities ($P_{X1...n}$) of getting selected. Competition in the market, therefore, is a condition where homogenous or slightly heterogeneous products in competition have more or less equal chances of getting chosen if not bought. The theory further assumes that competition is governed by three elements which can be called infrastructure for competition. These are: a) price-cost ratio which illustrates the minimal cost of transferring to another product; b) time ratio which illustrates the convenience of transferring to another product at very minimal time, and c) information ratio which provides the availability of information to select the other product. Competition, then, can be measured through Competition Index which can be defined as the determinant of all adjusted probabilities ($P'_{x1...n}$) that homogenous or slightly heterogeneous yet competing products ($X_{1...n}$) which are available in the market would have their chances of being selected (Gabriel 2014: 11).

The theory further assumes that the reason why prices of competing goods tend to be competitively the same is because buyers would have the convenience to switch to another competitive product. Convenience can be achieved if buyers could incur less time to switch to another competing product and would possess all the information they would need to make the switch. Sellers offering competing product would tend to have the same price if they are located close to each other but if one seller is situated at a distance, he may dictate a price of his own level since it would inconvenience a buyer to look for another seller and would condemn him to buy product he offers nonetheless.

From the principle of supply and demand, we could then determine why the price of the red onions had gone higher than the white ones at one time and had become more expensive than the other at other instances. The answer would lie on the mechanics of supply and demand. The red onions would have been cheaper due to the large supply of the onions without enough demand to meet the available supply. At the same time, the slide in prices of the white onions against the red ones would have also been due to supply and demand dynamics. A lot of the big white onions would have flooded the market that its price slid. But this is true if either commodity is a substitute of the other. If one kind of onion is sold at a high price, then buyers could easily switch to the other sold cheaper. But it may not be that simple. The white onions are imported while the red ones are locally grown. This means that the white onions are hauled in for domestic consumption through foreign currency as the instrument of exchange. We use our local currency in buying and selling goods in the local market but we need foreign currency to buy goods outside of our borders. Thus when a country's currency appreciates (rises in value relative to other currencies), the country's goods abroad become more expensive and foreign goods in that country become cheaper, holding domestic prices constant in the two countries. Conversely, when a country's currency depreciates, its goods abroad become cheaper and foreign goods in that country become more expensive (Mishkin 2001: 153).

> Anything that increases the demand for domestic goods relative to foreign goods tends to appreciate the domestic currency because domestic goods will continue to sell well even when the value of the domestic currency is higher. Similarly, anything that increases the demand for foreign goods relative to domestic goods tends to depreciate the domestic currency because domestic goods will continue to sell well only if the value of

the domestic currency is lower (Mishkin 2001: 157-158).

In this scenario, the value of domestic currency as against foreign currency would have an effect on the local and imported goods. At this point, we have to make it clear that the economic system in a territory where a certain local currency circulates is like a closed system where goods locally produced fulfill the local demand. The link of this domestic economy to another local economy in another territory would be the goods exchanged through foreign currency. A businessman who wishes to bring in goods produced from other territories would need to secure them through foreign currency from which he has bought or converted his local currency from. If there is a higher demand for domestic goods, then there will also be a higher demand for the local currency. But if the demand for imported goods soars, there will be a corresponding demand for foreign currency to purchase those goods and make them available in the local market.

> If a factor increases the demand for domestic goods relative to foreign goods, the domestic currency will appreciate, and if a factor decreases the relative demand for domestic goods, the domestic currency will depreciate (Mishkin 2001: 159).

This condition follows the Law of One Price. This law assumes that "if two countries produce identical goods, the price of the goods will be the same provided that transaction cost is low and there is very little barrier to trade." An application of this law is the Theory of Purchasing Power Parity which assumes that "exchange rates between any two currencies will adjust to reflect changes in the price levels of the two countries" (Mishkin 2001: 156). Let's say that a bag of cement costs 2 dollars in the United States while it costs 100 pesos in the Philippines. If the cement which is identical to both countries rises 10% in price in the Philippines and remains the same in price in the US, the exchange rate between the peso and the dollar will have to depreciate also by 10%, in order for the Theory of Purchasing Power Parity to hold. If the exchange rate runs at 50 pesos to the dollar, then the resulting rate would be 55 pesos to the dollar.

These rules, however, suffice long run conditions where demands for consumer goods locally produced compare with manufactured goods in other countries. There are other variables that would affect exchange rates in relation to goods demanded. A rise in the country's price level relative to the price level in another country causes its currency to depreciate. If foreign goods are flooding the local market, the price of local goods will sink for lack of demand and the price of foreign goods will rise due to high demand. Lack of demand for local goods with such low prices would create a falling profit for local producers but high demand and rising prices for foreign goods would escalate profit for foreign manufacturers. This would result in the depreciation of the local currency due to the high demand for foreign currency to purchase those goods and sell them in the local market. Thus increased demand for imports depreciates the local currency while increased export appreciates the local currency. If foreign goods however are flooding the local market, the government could impose taxes (tariff) on imported products or restrict the entry of incoming commodities at a certain level (quota). These protective measures could appreciate the local currency for it could limit competition with imported goods and maintain a high demand for local commodities. In the long run, if a country has high productivity, shipping more exports to suffice the market overseas and fulfilling the local demand as well, permits the local currency to appreciate (Mishkin 2001: 158).

Now if we look at the onions, we may find that the amount of export or import of onions could not influence the

The Milieu-already-constructed

fluctuation of the exchange rate unless onions are a large component of the economy that every kitchen should have them for a meal or onions are the country's major export. Exchange rates are actually determined through spot transactions, negotiated by banks in their local and foreign currency deposits, which account to millions and billions each day. Since these assets are deposits, then what entice banks to shift from local to foreign currency or vice versa, is the fact that one deposit would earn more than the other. But it is presumed that both currencies are a direct substitute of the other. An investor could shift possession of either currency depending on the profit he could gain from it. Which deposit would yield higher return depends on the expected appreciation or depreciation of the local currency against the foreign currency and the interest rate on the deposits.

For a local currency against the foreign currency, it would be presumed that at a ratio of 1:1, the local currency has equal value with that of its foreign counterpart. An appreciation of the local currency would mean a lower ratio for the local currency against the foreign currency; let us say 0.975:1 (0.975 for the local currency and 1 for the dollar). Thus with an interest of 3% on local currency deposits and an expected appreciation of the local currency by 2.5%, then the expected return on the local currency deposits relative to the foreign currency deposit would be 5.5% (3%+2.5%). On the other hand, if the local currency were expected to depreciate by 2.5% over a certain period of time, then with an interest of 3%, the expected return on the local currency deposit relative to the foreign currency would be 0.05% (3%-2.5%). The scheme could be written in this fashion (Mishkin 2001, 161):

$$R_l = i_l - ((E_b - E_a) / E_a)$$

The return in local currency (R_l) deposit is equal to the interest rate in local currency (i_l) added to the factor of appreciation or depreciation of the currency. On the same note, the return on foreign currency deposit would be computed in this manner:

$$R_f = i_f + ((E_b - E_a) / E_a)$$

The return on foreign currency deposit (R_f) would be equal to the interest rate (i_f) subtracted from the factor of exchange rate appreciation or depreciation. In this scenario, with the 2% interest rate on foreign deposits and with the same movement of exchange rate as before, the return on foreign currency deposit would amount to 7.1%. With the expected return on foreign currency deposit greater than the expected return on local currency as affected by the expected depreciation of the local money, investors would try to hold more of the foreign currency than its local counterpart. The relationship between the interest rate and the exchange rate could be plotted in a graph, the same way as the supply and demand could be analyzed in the same method (Figure 3.14).

With the exchange rate at Ea (Figure 4.15), the local currency is weaker than the foreign currency. The exchange rate levels off at E and appreciates at Eb. As the exchange rate moves at the upper region, investors would favor holding on to the foreign currency since it appreciates and as the exchange rate moves below the equilibrium, it would be more profitable to hold on to the local currency since the local currency appreciates. Now if the local currency depreciates further, more investors will try to procure more of the foreign currency. Since the supply of the foreign currency is limited, the demand for it increases to a level that the supply cannot anymore sustain it, raising its price according to the Law of

The Milieu-already-constructed

Supply and Demand. On the other hand, since there are no takers for the local currency, then its price decreases, and soon, the exchange rate will also level off at the equilibrium level.

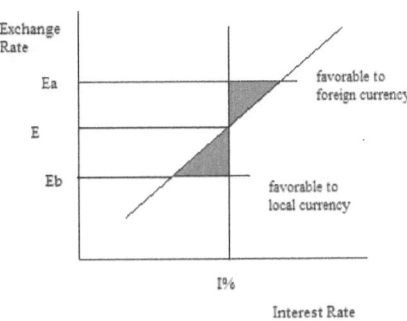

Figure 4.15 Regions Favorable to either Foreign or Local Currency

A foreign currency investor, therefore, determines profitability through his expectation of the future rise or decline of the value of the local currency relative to the foreign currency. Thus an investor would watch out for several indicators. One would be the interest rate. An increase in the interest rate would have an effect on the exchange rate.

An expected shift of the interest rate would result in the shift of the equilibrium ratio. This would result in a weaker local currency and businessmen would not like a high interest rate for it would make for a sluggish business climate, stifling expansion and taking away productivity. On the other hand, indicators such as expected increase in tariff, quotas of foreign goods, rise in productivity, increase export and lower import would shift the schedule of interest and exchange rate upward, resulting in an increased equilibrium and appreciating the local currency in return. On the other hand, a projected increase of the inflation rate would move the interest-exchange rate below the equilibrium, declining the value of the local currency (Mishkin 2001: 165-170).

Remember, all these projections and expectations are based on prevailing situations and future conditions. Thus banks, which engage in spot transactions, would behave either by buying or selling these currencies based on the indicators which, in return, would either appreciate or depreciate the local currency.

What happens now to our onions? Well since the white onions are imported, a depreciation of the local currency would increase its price since it would take more local currency to buy them, considering the Law of Supply and Demand and the fluctuation in the exchange rate. Thus the price of onions is set, with the Law of Supply and Demand, and the fluctuation in the exchange rate. While the price of the onions is determined by how much suppliers can fulfill the demand and while the value of money is ascertained relative to the strength or weakness of the foreign currency through bank transactions and foreign currency investments, the whole dynamics works through our expectations and projections of prices and profit which have been indicated by factors brought about by manufacturers who fulfill exported and locally demanded goods and how government measures are installed to curb inflation, maintain monetary supply and impose taxes.

There is just another thing to bear in mind. Even if economic theories deal with money and commodities and how they are exchanged, what economics explains is the behavior of people as to how they buy and sell commodities with corresponding quantities of money. Moreover, it is presumed that part of the *ceteris paribus* assumption of these theories, even the configuration of the demand and supply, is the postulate that buyers and sellers are imbued with perfect

The Milieu-already-constructed

information. It is presumed that the buyer has sufficient information to search and decide what commodity to buy and the sellers as well have enough information about the goods they are selling. Thus with the perfect information, they could choose and set the right price. But perfect information is hardly achievable. The quantities of goods and money with the information available, therefore, become a measure as to how buyers and sellers behave and interact to suffice certain desires or interests. The milieu of buying and selling in exchange for goods and services is constructed in a milieu already built within the contours of the Law of Supply and Demand. The market may be conceived in an already constructed milieu where prices are set through the dynamics of how much supply reaches the market and how much would be the demand for it with money as the means of quantification and exchange.

Postscript 1

A theory can be likened to building a prefabricated house. You have seen the model house, you bought one, the company disassembles the components of the house, shifts it to your location and reconstructs it. But before the reconstruction, at the moment of shipping the components, you already know how it would look since you have seen its model before. The theories' capability to deconstruct, reconstruct and foreconstruct phenomenon can be summarized in Table 4.1.

Reading the World and Inventing Science

Table 4.1 Summary of Theories in their Deconstructive, Reconstructive and Foreconstructive Mode in Milieu-already-constructed

Theories and Phenomena	Deconstruction of the phenomena into parts	Reconstruction of the parts into their relationship	Foreconstruction of what occurs
Newton's Law of Motion Mechanics at a speed within human grasp	Speed (S), Distance (D), Time (t), Force (F), Mass (M), Acceleration (a), Kinetic Energy (KE), Velocity (V)	Speed (S) is directly proportional to distance (D) but inversely proportional to time (t). Force (F) is directly proportional to mass (M) and the rate at which the body changes velocity (a). A body that moves needs an amount of energy (KE) which is directly proportional to the mass (M) of the body and to the square of its velocity (V).	When an object moves the speed is governed by the distance over the time it takes to cover the distance and will continue on moving with the force that sustains it until another force obstructs its movement. A body that moves needs or spends energy to move it.
Special theory of Relativity Mechanics at the speed of light or close to it	Speed of an object (S), Speed of light (C), Length (L), Mass (M), Time (T), Energy (M)	The speed (S) of a moving body is relative to another moving body. The speed light of (C) is the only constant. A body that	When a body moves, its movement is relative to another body in motion. (When you walk, your movement is dependent upon the earth that

The Milieu-already-constructed

Theories and Phenomena	Deconstruction of the phenomena into parts	Reconstruction of the parts into their relationship	Foreconstruction of what occurs
		moves contracts in length (L) and accumulates mass (M), but dilates time (T) in its movement. Mass (M) is convertible to energy (E) equal to the multiplication of mass to the square of the speed of light (C).	moves). Only the speed of light is not affected by any other movement. When a body moves it shortens in length in the direction of its velocity, it becomes heavier and the duration of time that it is experiencing is longer. At the atomic level, mass is readily convertible to energy without even moving it.
Cell Theory Constitution of all living organisms	Cell	The cell is the basic unit of life. It is the structure of living organisms. Cells come from their own kind.	All living things, whether plants or animals are composed of cells. The human body is a complex system made up of different kinds of cells. Plant or animal cells come from their own living kind.
Chromosome Theory Constitution of the inheritance carrier	Chromosome	The chromosome is the inheritance carrying unit of the cell.	The characteristics of plants and animals are passed on through the chromosome in the cell.

Reading the World and Inventing Science

Theories and Phenomena	Deconstruction of the phenomena into parts	Reconstruction of the parts into their relationship	Foreconstruction of what occurs
DNA Theory Composition of the inherence-carrying material	Deoxyribonucleic Acid	The inheritance carrying component in the chromosome is a complex molecule called DNA.	The information about the inherited characteristic is contained in the DNA. The organism's specific character is contained in the structure of the DNA.
Law of Supply and Demand Operation of the market	Price (P), Demand (D), Supply (S), Quantity (Q), Equilibrium (E)	Supply (S) behaves with quantity (Q) of goods being sold to be directly proportional to price (P) but demand (D) behaves with quantity (Q) of goods being bought to be inversely proportional to price (P). The intersection of the demand and supply is the equilibrium (E) price and quantity.	Sellers would produce and sell more when prices are high and buyers would buy more when prices are low. The agreement of buyers to buy and sellers to sell at a certain price pertaining to the volume of goods being transacted is the equilibrium.
Law of Diminishing Marginal Utility Consumer satisfaction	Marginal Utility (MU), Commodity (C)	An increment in satisfaction or marginal utility (MU) is achieved with the consumption of	You get satisfied with the consumption of one commodity and would like to consume an additional unit

The Milieu-already-constructed

Theories and Phenomena	Deconstruction of the phenomena into parts	Reconstruction of the parts into their relationship	Foreconstruction of what occurs
		one unit of a commodity (C), but would later diminish the marginal utility with the consumption of a few units more of the same commodity.	more, but would eventually get used to it, to the point that you get dissatisfied if you consume more of it.
Theory of Perfect Competition Producer's profit maximization	Average Revenue (AR), Marginal Cost (MC), Price (P), Quantity (Q)	Profit is maximized if a firm produces a quantity (Q) of products and sells it at a price (P) when the average revenue (AR) or the revenue derived per quantity sold is equal to the increment of cost of producing it or marginal cost (MC).	Producers would sell the commodity they produce at a certain volume and gain maximum profit when the addition of money they spend in producing it is the same as the revenue they would derive from the production of each commodity.
Theory of Time, Information and Money Competition in the market	Price (P), Time (T), Information (I), Competition (C)	Market competition (C) is achieved when a number of commodities of the same price (P) have equal chances of getting selected where consumers	Similar or quite dissimilar products compete with each other when buyers could easily switch to buy one commodity against another if it would incur them less time to

Theories and Phenomena	Deconstruction of the phenomena into parts	Reconstruction of the parts into their relationship	Foreconstruction of what occurs
		would conveniently have less time (T) to buy a homogenous product considering that they have all the information (I) to prefer the other.	switch and buyers have enough information to do so, resulting in the same price products are sold. Thus competing products sold close to each other would have the same price.
Law of One Price Prices of goods produced by two countries	Price (P), Goods (G), Countries (C)	The prices (P) of identical goods (G) produced by two countries (C) will be the same provided there is minimal transaction cost and barrier to trade.	The price of the same products sold and produced in the different countries will be the same.
Theory of Purchasing Power Parity Exchange Rate adjustment	Exchange Rate (ER), Price (P)	The exchange rate (ER) between two countries will adjust to reflect price (P) changes in the two countries	Exchange rate varies according to the prices of goods in demand and sold. Though identical goods should have the same price, price variations depend on the level of exchange rate.

    The preceding discussion would demonstrate how theories are able to construct and reconstruct the milieu for us. And this is the milieu-already-constructed. Theories are able to build and rebuild reality in a milieu that presents itself for discovery. With this, we are confronted with two questions, "what is reality" and "what are the outlines of this milieu-already-constructed?"

The problem of reality was first addressed by monists in the Pre-Socratic tradition, proposing that everything can be reduced into one entity. It is, in fact, a problem of how "the many" is related to "one" and "one" to "many." To bridge the gulf between the many and one, Pre-Socratic philosophers assumed that the "one" with which everything or "the many" is reducible into, is nothing but a substance. Thales said it was water, Anaximanes proposed it is air, Heraclitus argued it is fire, Xenophanes opted for both earth and water. It was Democritus who advocated that everything is reducible to an indivisible, "uncuttable" particle called atomos. (Miller 1992: 58-59).

When Plato came into the picture, his problematique changed. He tried to find the concept that would bridge the being and the becoming and not one with many. He argued that there is reality out of everything that appears. What appears changes. They become. But there is an unchanging, transcendental concept or being through which the becoming is caused. That is reality. This is his Theory of Transcendental Form. Instead of advocating a certain substance through which everything is reducible, he singled out the characteristics of this form or essence. He argued that the being or the nature of form is characterized with its objectivity (it exists outside of our mind or will), transcendence (it lies outside of time and space); external (it is not subject to motion or change); intelligible (it is comprehended by the intellect); archetypal (it models every kind of thing that exists or could exist); perfect (it contains all the elements of the nature of the thing it seeks to model) (Miller 1992: 76).

But this theory was attacked by Aristotle, his student. He claimed that the nature of things is not separate from the thing itself. The "whatness" or nature of things does not lie outside of time and space but it is inherent in the object itself. That is reality. And that is his Theory of Immanent Form. Since reality is in the object that appears, then such object has been caused. For this, Aristotle, proposed four causes that make up the essence of things. These are the material cause (that which make up the object); the efficient cause (that which creates the object); the final cause (that which accounts for the purpose of the object); and the formal cause (that which sums up the whatness of it) (Miller 1992: 90-92).

Reality is interpreted experience. We are not speaking here of truth, since truth still needs to be interpreted for us to understand it. But reality is interpreted experience. And who interprets reality but we. We experience an event or a thing and have interpreted it to be in existence. Thus it is real. But reality has many types.

In 1600, the rivalry between empiricist and rationalist was waged to determine the epistemic validity of our claim on things that we know. Knowledge is important but the problem is how do we know that the things that we know are valid. The empiricist argues that the claim to knowledge is verified through observation. Knowledge is achieved through the senses and we validate what we know through sense impressions. What we know, on the other hand, is validated by others sensing the same. Sugar is sweet because our sense of taste tells us it is so and we have a language for such a sensation – sweet. We know that fire is hot because the temperature produces sensation when we touch it and our term for that sensation is – hot. Immanuel Kant, termed this *a posteriori* knowledge or knowledge obtained purely from the senses (Stewart and Blocker 1996: 238). The kind of reality, therefore, that these cases present is nothing but *sensed reality*.

But we cannot fully rely on our senses. Descartes a 17th century philosopher and proponent of the rationalist school argued that the mind can also generate knowledge by intuition. He popularized the line "I think therefore I am." Other than our senses, the mind can generate and determine the validity of things that we know. Kant termed this *a priori* knowledge or knowledge obtained independently of our senses (Stewart and Blocker 1996: 238). For example, we know that one finger added to another finger makes two fingers. But if you have one million, three hundred forty six thousand beans added to another three thousand, four hundred sixty six peas you will have one million, six- hundred forty four hundred sixty six, beans and peas. You don't have to count the beans and peas to ascertain the claim because by operation taken from the law of addition, the sum is valid. If you multiply 2^2 with 2^3, the answer would be 2^5. The answer can be verified through the Law of Exponents. And it can be validated by the same operation. We have a term for it and the process on how it operates. Knowledge such as this is abundant in mathematics and science (Durant 1961: 266-267). These realities can be termed *axiomatic realities*.

But we cannot limit ourselves with only these two realities. Human beings are social and interactive beings. When two people communicate with each other, they create or reinforce certain meanings or rules that govern their interaction. They could share facts but as they share them, they exchange interpretations and meanings. These meanings are not visible neither are they generated by universally accepted principles or laws. But they are produced and reproduced through interaction (Taylor 1977: 51). Moreover, as human beings interact, they either create or reinforce rules or structures which either constrain or enable their actions

(Giddens1984: 23-25). When we address someone senior to us, we attach sir or madam to show our politeness. Thus we are constrained by the social structure not to utter the word and enable us to say it. Furthermore, individuals having reached a certain degree of communicative rationality could arrive at a consensus (Roderick 1986: 113). People who engage in negotiations or indulge in argumentative speech could arrive at an agreement. Thus, the meanings, social rules or structures and consensus are realities which can be termed ***interacted realities***.

Another reality is the about our interaction with the text. Separate in its nature from interacted realities, this type of reality can be termed ***textual reality*** which will be discussed fully in Chapter 5.

All these are realities which are, in fact, interpreted experiences. Even if we see the color red, what strikes the retina or our eyes are actually wavelengths with 6,563 Angstrom units and one Angstrom is as small as 1/1,000,000,000 or one over one billionth of a meter. And we have a term for that sensation. We call it red. Likewise, being hot is actually temperature at certain degrees at the higher end of the thermometer. And we have generated a term for that sensation. We call it hot when we feel it. These are sensed realities. But being hot or the color red is an interpretation of the wavelengths and temperature of the things we sense. Moreover, arithmetic operations, geometric theorems, scientific equations, are outright interpretations which the mind can generate and can be used to validate certain mathematical or scientific claims. On the other hand, the agreements that we produce, the rules that we device and the meanings that we share in our interactions are real. It does not matter, at this point, about the truth of the interaction. Interacted realities are observable in its operation. The

The Milieu-already-constructed

interaction can be observed. It can even be counted. All these realities, however, can be captured and textualized.

This brings us to the question of "what the milieu is" and "what the milieu-already-constructed is?"

The world is an aggregate of realities. And since realities are interpreted experiences, the world then is an experience. It is an experienced reality. The milieu-already-constructed then is a discoverable, pre-fashioned pattern of sensed and axiomatic realities that impinge on our interaction. Due to the fact that the milieu-already-constructed is a milieu already made then the theories can textualize it by capturing its patterns and presenting it through universalized principles and quantification. Such quantification can easily exhibit itself through measurements. These patterns are even isolated from the person who observed, sensed, or textualized these patterns which can perpetuate themselves through governing principles. How the capturing of patterns and universalizing them to principles are done is by closure, then the theory disassembles these patterns and assembles them back again. The deconstructive-reconstructive-foreconstructive capability of theories sets it apart from any other text.

Since the milieu-already-constructed is discoverable, how about those phenomena which we have not yet discovered or have not yet experienced? But before we intricately delve into this puzzle, let us first look at the outlines of the milieu-under-construction.

Chapter 5
The Milieu-under-construction

It is but logical to assume that the milieu that presents itself for discovery is a pre-existing milieu, for there is nothing to discover if it does not exist before hand. But the beauty of how this milieu operates is matched by the continuous building and re-building of the milieu we can call the milieu-under-construction. This is a milieu that is continually being built. And with it is its nature of uncertainty. Social action can easily be presumed to be the sole builders of this milieu. But the physical and biological worlds are not exempt from this milieu that is continually being built despite its nature of existing in a milieu-already-constructed. What theory does for the milieu that is already built is also the same deconstruction, reconstruction and foreconstruction of the milieu that is being built and rebuilt.

Textualizing Quantum Phenomenon

There are three fundamental factors ingrained in the subject of mechanics: the location in space, the duration in time the object has been located and the combination of these two factors as the object shifts its location at a certain duration -- speed or velocity if it were going at a defined direction. Matter occupies space and has mass. Space is the locale which the body that has mass occupies. The history of a particle or an object in regard to mechanics can be determined as regards its location in time and how fast or slow it is moving towards a certain direction. For the classical Newtonian mechanics, speed is obtained by dividing distance over time. The faster the speed, the longer the distance covered but the shorter it would take to travel that distance. One can take a meter stick, measure the distance and take a stopwatch to mark the time. But the Newtonian concept has philosophical consequences. One would always presume that the person measuring the body in movement is at rest or zero motion.

If you are holding the stopwatch, you can very well feel you are not moving while the body you are measuring does. Even the road where the body moved is resumed to be not moving. But Einstein challenged Newtonian mechanics when he proposed the Theory of Relativity. Einsteinian mechanics assumes that motion is relative to another body in motion. If a man is measuring the speed of a sprinter on the tracks, then the man at the end of the finish line may be presumed to be not moving, but relativity says he is, even the tracks which they are standing on. The real and philosophical consequences of this theory, however, are great. This means that there is no absolute rest. Without an absolute rest, then speed is relative. That would also mean that there is no absolute space and since speed is relative then time is also relative to a body in motion. Everything would then be in motion and even space and time would be relative to motion. This would mean that space would contract and time would lengthen as the body moves. And since all matter has mass, then mass would increase in terms of its motion. If everything is in motion, then there ought to be something which is constant. That constant is the speed of light. It is constant even

The Milieu-under-construction

if its source is moving. Since the speed of light at 186,000 miles per second is also unmatchable, then this speed can be used as a measuring device to ascertain speed, space and time.

It would take just simple mathematics to do the conversion. Let's say 1 mile of distance would take 0.0000056 second for light to travel. Then 1 mile or light-mile is equal to 0.0000056 second. In the same manner 1 second or light-second would be equivalent to 186,000 miles or light would cover a distance of 186,000 miles in one second. In this configuration, therefore, space and time are no more separate and the speed of light has united them. If a body in space can be located in space using three points (X,Y,Z) in a three-dimensional axes, then time can be incorporated to account for four-dimensional plot. This would mean that everything moves about in space and time or space-time. A person who is lying in bed may be at rest but, in fact, he is moving in time and if plotted along the four-dimensional axes, he is moving on the time axis. An object in space-time can be plotted, creating a world-line.

The person (Figure 5.1) who is sitting on a chair and working on her table may be said to be not moving in space since she is simply occupying a certain locale but she is, in fact, moving in the duration of time since relativity treats time to have the same quantifiable units as space. As she leaves the chair and jogs, she then garners space and time. The line she draws in space-time is her world-line which is her history in regard to Einstein's mechanics.

Einstein's mechanics is highly explanatory of speeds close to the speed of light and its consequences to space, time and mass as the body moves. It is able to explain bodies in vast distances. Moreover, we could still locate the body in space-time and account for its history along its world-line. But as the theory accounts for vast distances, it breaks down, however, on the very small. An even stranger phenomenon occurs on the scale of the very minute.

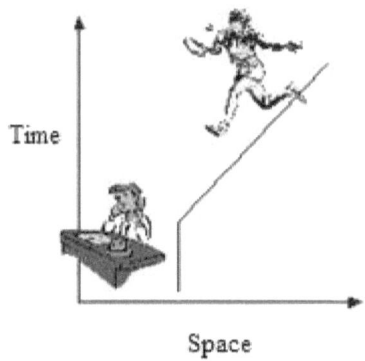

Figure 5.1 World-line of a Person in Space-Time

At the time when Einstein wrote on the relative properties of motion with his paper "On the Electrodynamics of Moving Bodies" in 1905, another branch of physics was evolving separately from his proposition on relativity. Another revolution was taking place. And this began with the bugging problem of radiation.

We can illustrate this with an example. After a long drive, we park our car. If we open the hood and touch the engine, naturally we feel it warm, if not hot. The casing of the engine may not be hot at first for it is inside each cylinder where fuel is burned from 600°F at compression to several thousands of degrees during ignition. The manner by which heat reaches our hand through direct contact with the source is called conduction. Heat is passed along through the vibration of molecules and free electrons from the cylinder to the engine casing, onto our hands in direct contact with it. But if we wave our hand a little above the engine still we feel it warm. Heat then is transferred via convection. It is the manner by which heat is transmitted through the actual movement of hot materials. In this instance, it is the upward swell of hot air above the engine that reaches our hand. But if we close the hood and touch it instead, we feel it warm as well. This time, heat is relayed neither through conduction nor convection anymore, but through radiation. Radiation is a process by which heat or energy is transferred through the emission of electromagnetic waves. But a big problem here exists.

Since 1850s, it was understood that light is a form of electromagnetic wave as proposed by James Clerk Maxwell. And the alternation of electric and magnetic waves at perpendicular orientation forms various vibrations depending on the wavelength and frequency of the wave. It would then follow that we could designate frequencies as high as they get, making the transmission of energy infinite. For a physicist, this is incredible. Infinity defies us to set parameters even denies scientists of their ability to accurately calculate and measure. Moreover, since the transfer of energy is in the form of electromagnetic wave, then the conveyance of energy would have to be smooth, from low frequencies to higher ones. But experiments run contrary to these.

If a piece of metal is subjected to direct heat, it glows red hot first and as temperature increases, it turns white hot. In a spectrum, we know that red has lower frequency than violet, below red is infrared and higher than violet is ultraviolet. But why is it that energy is transferred at red and skips immediately to white since white contains all the colors or frequencies in a spectrum.

In early 1900, Germany was witness to some of the precise experiments on radiation. During a party which Max Planck and his wife hosted, Heinrich Rubens, his colleague, showed him his latest findings on black body radiation (Kaku and Thompson 1987: 38). A black body is a material which absorbs radiation effectively. A body that absorbs radiation well transmits it efficiently just the same. A black shirt feels warmer once worn under sunlight than a white shirt. A black nail will turn red hot faster than a shiny one. This is because a good reflector is a poor radiator. Thus a thermos flask is coated with shiny silver inside so it will not radiate heat fast and preserve the temperature of the water it contains. But just as Rubens informed Planck, radiation it seems does not proceed in a smooth manner.

After the discussion with Rubens during that party, Planck went on to work out a smart solution to the problem. The trick was not experimental but mathematical. He devised a new equation to fit in Ruben's data and proposed that energy is not transmitted in wavelike fashion, just as scientists thought, but in packets called quanta. The size of each packet is so small. It is determined by a number h= 6.5×10^{-27} erg-seconds called Planck's constant (h) multiplied by the frequency (f) of the wave to determine the amount of energy (E) the electromagnetic transmission possesses.

$$E = hf$$

While light is a wave, the energy it contains is not transmitted in the form of a wave but in granules or quanta. The energy swings from one whole packet to the other. When one heats a nail over a flame, most of the radiant energy initially emitted at 300°C possess longer wavelength at lesser frequency compared to red light. This is the infrared range. At 800°C the nail becomes luminous and appears visibly red hot. And at 3,000°C the temperature of a heated filament in an incandescent bulb, the nail contains enough short wavelength oscillation that it appears white hot (Sears et. al. 1979: 273-274).

Planck was cautions of his findings even when he presented it to the Berlin Physical Society. But it took Einstein in 1905 to push Planck's Quantum Theory a little farther when he wrote his own Theory of Photoelectric Effect. Physicists, for a while, thought Planck's theory of chopping down light into granules, which behave like particles, was preposterous

The Milieu-under-construction

but Einstein proved otherwise. He theorized that if light were granular in nature then it could displace some electrons from a metal plate once light strikes onto it. Metals have sufficient loose electrons on its outer energy level which permit it to be good conductors of heat and electricity. The displacement of electrons would cause an imbalance and generate electric current. Einstein wrote down his own equational interpretation of the phenomenon and it did not take long for scientists to confirm both Planck and Einstein to be correct.

But this is not the only implication of Planck's quantum revolution. It was already discovered that if one electron becomes excited by heat or some other means it oscillates from a region of high energy (position toward the nucleus) to low energy level (location away form the nucleus), certain amount of photon or quanta of light is produced. If the electron of heavy element, which possesses a great number of protons producing high positive charge, is knocked out from the region very near the nucleus, the vacancy which another electron fills up causes a certain quanta of radiation in the X-ray range. In this case, the production of energy at the quantum level is always accompanied by one thing – motion or mechanics. Remember, if we were talking of mechanics, location and velocity are important.

An electron, however, is not stationary. It moves around the nucleus. But its manner of movement was theorized before to be a well-defined orbit like that of the planets that revolve around the sun. Moreover, the electron is so small that we could not even view it under the microscope. If a packet or quantum of light is minute enough, an atom, more so an electron is even smaller. Thus a microscope, which needs light to make its subject luminous, will not be sufficient to make an electron visible in the human eye. The only way to observe an electron and to ascertain its velocity and position around the nucleus is to beam electromagnetic waves onto it. How the wave will hit the electron and bounce off will give an indication of its position and velocity. But this preconception never came close to presupposed reality.

In 1927, Werner Heisenberg found out that as one beams electromagnetic rays into an atom, the rays as measuring devise disturb the electron and ascertaining its position and velocity at the same time becomes impossible. This would mean that one could know the location of the particle but not its velocity. And if one has determined how fast it is going he would not know where it is exactly. The disturbance caused by the measuring instrument becomes part of particle's nature itself.

In Newtonian mechanics, however, deriving the velocity and position of matter is not a problem. If we roll a ball on the floor, we can measure how fast it has moved using a meter stick and a stopwatch. And as it rolls, we could observe where it is traveling. In Einstein's Theory of Relativity, matter would have a designated path in space-time which can be drawn through its world-line. Matter can be plotted in a graph of space and time and can be located in the space-time coordinate. But minute particles seem to have a world of their own.

We cannot ascertain their position and velocity at the same time. We could know one or the other but not both at the same time. This is Heisenberg's Uncertainty Principle which every experiment faithfully confirms. Heisenberg's equation states that when the uncertainty of the particle's position (Δq) and the uncertainty of its momentum (Δp) are multiplied, the product approximates Planck's constant (h) (Ridley 2001, 144).

$$\Delta q \cdot \Delta p \geq h/4\pi$$

Heisenberg postulated that the uncertainty is not an issue of measurement of the device used to measure it but this behavior is really part of nature itself at the level of the very small. At this point, it seems nature has gone mad. If we could calculate how fast a particle is moving, we could not ascertain where it is, and if we are able to locate it, we wouldn't know where it is going.

In 1924, Louis de Broglie extended the wave-particle duality of light to matter itself. He postulated that all matter in motion has wave properties. If we throw a ball in mid-air, the momentum (p) of the body which is determined by the mass (m) of the body multiplied by its velocity (v) creates waves whose wavelength (λ) would be equal to Planck's constant divided by its momentum.

$$\lambda = h / p$$

The wave becomes a property of the ball in motion. Since an electron also moves around the nucleus, then the electron will also have its wave properties whose wavelength may depend upon its velocity. Thus just as a particle behaves both as wave and granule, an electron would then be both wave and particle just the same. The Viennese physicist Erwin Schrodinger took this theory farther. Building on de Broglie's particle-wave concept of matter and Heisenberg's Uncertainty Principle, Schrodinger proposed that an electron is not anymore a point particle but a spread out entity. He developed his formula known as Schrodinger wave equation, describing how this wave obeys. This shatters the previous notion that the electron circles around the nucleus in a designated orbit. Since it is both wave and particle, then it would appear like a fuzz or a cloud around the nucleus and since we would not know exactly where it is, then probability operates to give us a picture of the region where it could be located. From the deterministic properties of large bodies, the minute particle seems to behave in a probabilistic manner.

Quantum mechanics has taken our imagination far beyond and forces scientists to recast their philosophy of science to determine how everything began. Even Einstein could not accept the revolutionary propositions of quantum mechanics through he was partly responsible for originating it by writing down his Theory of Photoelectric Effect. He was then forced to argue, "God does not play dice."

Nature has gone crazy. We cannot know where the particle is though we can determine how fast it is going and if it can be located, we cannot know how fast it is moving. Heisenberg assumes that this is really how nature behaves at the quantum level and this is not simply the result of observational limitation where measurement disturbs the objects being observed where such disturbance becomes part of the observation. Einstein rejects the generalization advanced by quantum physicists on the proposition that nature behaves probabilistically at this level. Einstein does not object to their findings since the results have conformed to experiments but Einstein argues that the theory is still incomplete. Einstein believes that there are still some deterministic laws waiting to be discovered to account for the behavior of the very small, unlike the probabilistic end that quantum mechanics advances (Kaku and Thompson 1995: 48).

But we can't just reject what the findings of the experiment present to us, that "we can't be certain of a particle's locality and velocity at the same time." Thus far, as we beam electromagnetic waves to an electron to ascertain where it is and measure how fast it is going; the more accurate we want to be, the higher the frequency of the wave we use and the greater the disturbance of the particle. Thus the observer becomes part of the observed and the result is a world of the observer and the observed significantly reconstructed.

Textualizing Life's Variations

Even if life is universally constructed with polymeric molecules of deoxyribonucleic acid (DNA), the information-carrying substance has the capability to form inexhaustible, permutational combinations that reproduce the variedness of life. Come to think of it, plants, animals, bacteria and viruses have DNA as the common information-carrying medium. But there are more than a million organisms and each organism may fall under different species but they are made up of the same genetic material with four nitrogenous bases to encode the kind of species they will be. Thus with four nitrogenous bases, an immensely voluminous number of combinations are generated to instruct cells to develop, replicate, maintain and organize themselves. Even life begins from its parent organism with only one cell that will soon divide to numerous cells and make up tissues, organs and species as a whole but that one cell carries all the information for one specie to live, grow and maintain itself. Thus even if life is constructed with the same genetic material that make up for an already constructed milieu for these species, the permutational combinations of nitrogenous bases of the DNA molecule creates various pieces of information in order to produce and reproduce various species in a milieu that is under construction.

In 1856, the Augustinian friar Gregor Mendel started out by collecting 34 distinct varieties of garden pea which was capable of self-pollination and fertilization. After sorting out the peas into certain traits like pod color (white, colored), and pod shape (round, wrinkled), he planted the seeds and artificially cross-pollinated sets of pea plants with distinct but opposite characteristics – white with colored; round with wrinkled. In each case, the hybrid offspring of the first filial generation F1 yielded only one trait of either the two parents. For those crossed between white and colored (Figure 5.2), the peas produced were only white, while for those crossed between round and wrinkled, the offspring yielded only round.

Where: A=White, a=colored		
F1:	White AA White Aa	x Colored Aa White Aa

Figure 5.2 First Filial Generation of Crossed Peas

Mendel called the parental trait that showed in the offspring as dominant while the trait that did not manifest in the offspring as recessive. Mendel called this the Law of Dominance. Then he allowed F1 hybrids to self-fertilize and the second filial generation F2 yielded ¾ of the dominant trait and ¼ of the recessive characteristic. After successive tries with the same outcome, he formulated the Law of Segregation which states that hybrids (cross-breeds) with one pair having dominant traits produce one half of the total offspring with qualities of the parent hybrid while another half has equal share of dominant and recessive traits (Weaver and Hedrick 1991).

For example, what would happen if you cross round and wrinkled peas? The result would be in Figure 5.3.

The Milieu-under-construction

	Where: A = Round (dominant), a = wrinkled (recessive)		
F1:	Round, Round AA	x	wrinkled, wrinkled Aa
	Round, wrinkled Aa	x	Round, wrinkled Aa

(only the round trait shows because it is dominant)

F2:	1/4 AA Round, Round	1/4 Aa Round, wrinkled
	1/4 aA wrinkled, Round	1/4 aa wrinkled, wrinkled

Figure 5.3 First and Second Filial Generations of Crossed Peas

Of the total population of peas produced in the crosses, 3/4 would exhibit the roundness of seeds while 1/4 would show wrinkledness. It was then presumed that an offspring receives two kinds of reproductive cells coming from each parent combined at random. Mendel then tried to confirm this assumption by crossing a dominant hybrid Aa with mainly recessive aa. The result was a population composed of ½ AA and ½ aa. This confirms that a hybrid is made up of two kinds of reproductive cells or gametes. The general rule was then stated that the segregation of each pair of differing traits from a hybrid was due to the presence of trait carrying elements called genes in the male and female reproductive cells with their own distinct qualities that unite at random during fertilization. Mendel assumed that these genes could exist in different forms called alleles. One gene could carry the seed color which may be dominant if it exhibits itself over the recessive trait. This assumption of the gene as carrier of hereditary trait underlies the idea of inheritance among living species and creates the foundation of genetics.

So far, Mendel's experiments and generalizations concern only those with opposite traits. Mendel proceeded to carry on the experiments in which parents differ in two traits such as shape and color. For example pea plants with round-yellow seeds combined with wrinkled-green seeds. From these crosses, Mendel formulated his Law of Independent Assortment which states that the behavior of differing traits in a hybrid combination is independent of all other differences in the parent plant (Weaver and Hedrick 1991). If two pairs of traits represented by A and a (round and wrinkled) and by B and b (yellow and green), then the double hybrids would be AaBb. This would form four kinds of gametes which are independent of each other, AB, Ab, aB and ab in equal proportion. These four gametes will unite by chance in fertilization (Figure 5.4).

Where: A,B = dominant but different traits a,b = recessive but different traits			
	AABB	x	aabb
F1:	AaBb	x	AaBb

The Milieu-under-construction

F2:

	AB	Ab	aB	ab
AB	* AABB	* AAbB	* aABB	* aAbB
Ab	* AABb	** AAbb	* aABb	** aAbb
aB	* AaBB	* AabB	*** aaBB	*** aabB
ab	* AaBb	** Aabb	*** aaBb	**** aabb

9/16 both dominant (AB)*, 3/16 one dominant (Ab)**, 3/16 the other trait dominant (aB)***, 1/16 both recessive (ab)****

Figure 5.4 First and Second Generations of Double Hybrids

For example, what happens if you cross-breed round-yellow pea with wrinkled-green pea (Figure 5.5)

Where: A = Round, B = Yellow (both dominant traits) A= wrinkled, b = green (both recessive traits)			
F1:	Round, Yellow AABB	x	wrinkled, green aabb
	Round, Yellow AaBb	x	Round, Yellow AaBb

F2:	AB Round- Yellow	Ab Round- green	aB wrinkled- Yellow	ab wrinkled- green
AB Round- Yellow	1/16 AABB Round- Yellow	1/16 AAbB Round- Yellow	1/16 aABB Round- Yellow	1/16 aAbB Round- Yellow
Ab Round- green	1/16 AABb Round- Yellow	1/16 AAbb Round- green	1/16 aABb Round- Yellow	1/16 aAbb Round- green
aB wrinkled- Yellow	1/16 AaBB Round- Yellow	1/16 AabB Round- Yellow	1/16 aaBB wrinkled- Yellow	1/16 aabB wrinkled- Yellow
ab wrinkled- green	1/16 AaBb Round- Yellow	1/16 Aabb Round- green	1/16 aaBb wrinkled- Yellow	1/16 aabb wrinkled- green

Figure 5.5 First and Second Generations of Double Hybrid Peas

In these successive crosses, 9/16 would be round-yellow seeds, 3/16 would be round-green peas, 3/16 would be wrinkled-yellow seeds, and 1/16 would be wrinkled-green peas. To confirm his findings, Mendel combined two hybrids having both dominant (AABB) and both recessive (aabb) traits. The result where four characteristic traits AB, Ab, aB, ab which reverted back to the four kinds of gametes formerly theorized.

Mendel's Theory of Heredity progressed without thorough knowledge of the basic unit of life. Mendel's work began in 1856 and continued on through 1863. He published his work in 1866, which got shelved for nearly 34 years. His proposition that genes carry these traits onto to the next generation made a breakthrough even without the knowledge of the cell. How are these traits passed on became the heart of another theory that could account for both the structure and mechanics of hereditary transmission.

Life always has its wonders. The Chromosome Theory, as discussed in the previous chapter, assumes that the chromosomes are the carriers of hereditary traits which an organism passes on to its offspring. These chromosomes are located in the nucleus of the cell. In order for the organism to grow, maintain and organize its functions, two things have to be satisfied. First, cells should reproduce and second, the chromosomes contained in each cell should be replicated in order to consistently maintain the transmission of hereditary traits and regulate its metabolism. The information-carrying substance should be contained in each reproduced cell. Amazingly, on the other hand, the cell is like a self-functioning machine. At the right condition, the cell divides and replicates itself through the process of mitosis. This process has been observed under the microscope and four phases have been assigned to characterize the progression. We have to bear in mind that the nucleus is the center of activity and the chromosomes which need to divide equally are to be watched in the process.

The interphase is the restoring stage where chromosomes appear to be long threads. Recall the fried sunny side up egg, the yolk can be considered the nucleus. Get some fine pieces of thread and drop them in the yolk. These threads can be likened to chromosomes. Now add a whole pepper inside the yolk. This is the nucleolus. We can visualize the cell to look like this in the interphase stage.

Now the first stage of mitosis is the prophase. At this stage, the chromosomes that appear like long threads thicken and shorten. Remember, that in the Molecular Theory of Genetics, the chromosomes are actually made up of a double-helix of deoxyribonucleic acid. It is actually a double coil of deoxyribose where the sugar content is linked together by phosphoric acid. In between these strands, the molecules of nitrogenous bases, purine (adenine and guanine) and pyrimidine (thymine and cytosine) are connected. Now how do these strands thicken and tighten. During the prophase a significant event happens. A spherical zone or centromere forms somewhere at the central region of the chromosome as it thickens. The centromere, however, is the site where the chromosome separates later into two. Now how can we demonstrate the emergence of centromere? Hold the rolled handkerchief and tie a knot somewhere in its length. The knot serves as the centromere. As the chromosome creates a centromere it forms a synapse or a pair. Synapse can be illustrated in this manner. Get a pair of scissors. Be sure that the handkerchief is not so tightly twisted. With the scissors, cut the handkerchief from one end until it reaches the knot. Do the same with the other end. Now you have an exact copy of the handkerchief which simply needs separation through the knot. The synapse is crucial in cell division for here is where the DNA duplication takes place. Remember our presumption that the resulting cell in this process should still contain the identical copy of the chromosome which bears the exact same information needed for the cell to perpetuate. It is during this process that the DNA is replicated. This will be discussed later. At this stage, however, another important event happens. The nucleus gradually disappears, the nuclear membrane, the substance that encloses the nucleus (or the elastic material around the egg yolk) forms spindle on the opposite side of the nucleus.

> The "fibers" or strands of the spindle may be formed by rearrangement of both nucleoplasmic and cytoplasmic materials, and are probably composed of protein chains, lipoproteins and a small amount of RNA in the form of a highly elastic gel (Burns 1972, 47).

At this time, the dividing cell is prepared for the next stage, the metaphase. During this stage, the paired chromosomes align themselves at the equatorial region of the

The Milieu-under-construction

nucleus or at that spot where the nucleus and the whole cell will split. The chromosomes, however, are arranged in such a way that the exact copy (homolog) of either chromosome lies on the region where the cell will separate, each identical chromosome converging on the opposite pole. Recall the handkerchief that we cut on both ends up to the knot. Since each side is identical, one side faces one pole and the other side faces the other. On the next stage which is the anaphase, the spindle fibers pull the paired chromosomes that the homologs split at the centromere. During the telophase, the chromosomes have reached the opposite poles and the equatorial region develops a cleavage where the entire cell separates and another identical cell is reproduced.

Now let's go back to the prophase stage where the most important event of cell division which is DNA replication takes place. It has been observed that as chromosomes thicken and shorten, the centromere emerges and synapse occurs, resulting in an exact, carbon-copy of homologous pair. We have demonstrated this by tying a knot on a rolled and twisted handkerchief and cutting both ends through the knot. But how does this cutting take place?

We have to bear in mind that chromosomes and the process of cell division have been directly observed under the microscope. But the cell being transparent would leave no trace even if peered through a powerful microscope. It needs to be dyed to permit its structural components to absorb the color and make it visible for observation. The process, however, kills the cell and stops its activity. In fact, the words chromosomes, chromatin (the material that makes it up) and chromatid (the arms or longitudinal half of a synapsed chromosome) have the same root word "chroma" which means an object that absorbs color. (Gamow 1961: 239). In this way the cell is observed but the DNA in its replication stages are too small to be detected under the microscope. Only the homologous pair of chromosomes that have taken on dye are visible. In order to account for this vital procedure, the Strand Separation Theory was proposed by A. Kornberg and his colleagues in 1960s in conformity with the molecular model.

> Kornberg has suggested... under the strand separation theory (that)... the two nucleotide strands separate by breakage of the hydrogen bonding between purine-pyrimidine pairs so that the molecule begins "unzipping" into a Y-shape figure. At the same time an enzymatic addition of complementary nucleotides (which must be available in the nucleus) by phosphodiester bonds takes place. The enzyme responsible... is DNA polymerase. It acts by adding a 5'-deoxyribonucleotide triphosphate to the 3'-hydroxyl end of the primer strand (Burns 1972: 285).

Unless your pants are button fly, this process is quite difficult to imagine. But remember how your zipper works. As you pull the small tag, the two rows of teeth separate into two. Now this is just as far as our imagination could take us for this action could not be directly observed. Armed with the molecular framework of the genes, what Kornberg and his colleagues did was to create a chemical synthesis of the DNA. Kornberg first succeeded in synthesizing DNA outside of its host from the DNA of *Escherichia coli* and the cellular enzyme DNA polymerase. But this resulting substance was biologically inactive. Then in 1967 Kornberg and his associates made a significant feat by using the single-stranded DNA of bacteria φX174 and DNA polymerase as the joining enzyme to create an active replicated DNA. Now it has been known that the bacteria φX174 with its single-stranded DNA only becomes double-stranded after infection on a host cell as a prelude to its replication. But Kornberg's experiment produced an active genetic material with the single DNA strand of φX174, producing 6,000 nucleotides as if infecting a host cell (Burns 1972: 285) The experiment, likewise, was a solid proof that the enzyme DNA polymerase was responsible for replication which serves as the other backbone of the DNA that is split in two.

But it should be noticed that the Strand Separation Theory only accounts for the replication of the DNA. Still, the issue of how the separation takes place has not been answered. Kornberg's experiment started out with only half of the DNA strand. Since the DNA is a double helix and it swivels, how does the separation occur? How does it unwind? How does the zipper unzip? What does the pair of scissors represent? Yet Kornberg's experiment was a hallmark in the understanding of how these genetic materials replicate and reproduce themselves.

If mitosis were a process to reproduce the cell with the replication of the genetic material, meiosis is a cellular process which is peculiar only to gametes, sex cells or reproductive cells of sexually reproducing organisms. It is a fact that organisms have even number of chromosomes. The fruit fly *Drosophila melongaster* has 8 chromosomes in each cell. Humans have 46. But only half of the complete set of chromosomes is contained in these peculiar gametes. Thus the sperm cell for the human male has 23 chromosomes while the egg cell for the human female also has 23 chromosomes. Yet while the gametes carry only half of the original number of chromosomes, it should satisfy the condition that an organism which possesses these gametes is a product of the equal number of chromosomes which the organism's parents shared to make him come about.

Now, how could this process be accomplished with ½ of the total number of chromosomes for the male parent added to another ½ of the female parent resulting also in ½ of the total number of chromosomes in the offspring? But this is true only to the gametes or sex cells which will combine through fertilization in order to complete the total number of chromosome for the ordinary cell. This is the reason why the gametes from each parent should carry only half of the total number of chromosomes. Here meiosis has a unique way of preserving half of the chromosome while equally taking in the chromosomes shared by both parents.

Under meiosis, the cell undergoes double prophase, metaphase, anaphase and telophase stages. Just like mitosis, the process begins with the interphase stage with the cell bearing the total number of chromosomes contributed by both parents. The spindle-like chromosomes begin to shorten and thicken at first prophase stage and a centromere appears along the chromosomes. Then the synapse begins to form, producing a carbon-copy of the chromosomes that would separate later. But unlike mitosis, an important event takes place at this stage. The chromatid of some chromosomes could crossover and recombine with another chromosome. This occurs when the spindle fibers form at the same time the nuclear membrane loses its distinction, the homologous chromosomes then aligns at the middle of the nucleus during the first metaphase. But strange things happen here. Instead of the chromosomes splitting at the centromere, the chromosomes are pulled on each pole without separating during the first anaphase stage. Thus at the first telophase stage, two daughter cells with half the total pairs of chromosomes have been created. But it does not stop here. The second prophase proceeds with the synapsed chromosomes aligning again at the mid-plate of the cell with each pair located at each pole. Then the second metaphase comes with the centromere breaking and splitting the chromosome in two and each pole receiving half of the pair. At the second anaphase the halved chromosomes are pulled apart, resulting in four daughter cells with half of the total number of chromosomes at the second telophase stage.

Now how does the crossing over occur? Picture two people with a complete set of legs and arms. The arms and legs serve as the chromatids. Crossing over is simply like

The Milieu-under-construction

crossing your arms or legs with the other fellow's arms or legs. The only difference is that, for the chromosomes, part of your arm becomes the other fellow's arm and part of his arm turns out to be yours. Before we move on to the details, recall the Mendelin experiment when he cross-pollinated peas with opposite characteristics. One characteristic showed up being the dominant trait. The other trait that did not exhibit itself in the offspring was considered recessive. But when he self-fertilized this generation, it generated 3/4 with dominant traits and ¼ with recessive characteristics.

Chromosomal evidence had to match these results since Chromosomal Theory presupposes that hereditary traits are contained in the chromosomes. Evidence of this occurrence, however, can be deduced during the crossing over stage at the meiotic prophase period. After the chromosomes had paired along the centromere, the chromatid of one chromosome links up and interchanges part of its chromatid with the chromatid of another chromosome, exchanging the traits they carry in the process. Then at the second metaphase after the chromosomes had split along the centromere, four daughter cells are produced, each with ½ the number of chromosomes.

Let's simplify the illustration. Suppose a cell only has 2 chromosomes, 1 from the male parent and another 1 from the female parent. During the meiotic process, the chromosome synapses, producing an exact copy of each chromosome along the centromere. If we use the handkerchief again as the illustration, we find the two handkerchiefs having a knot somewhere along each strand and each handkerchief having been cut along its length though the knot with four loose bands. One band from one handkerchief could crossover the other band of the other handkerchief, exchanging their parts. At this stage a crossover has taken place. After the crossover and recombination, the chromosome splits and subsequently separates, resulting in a chromosomal pattern having generally 4 types of chromosomal properties, 1 trait coming purely from one parent, the other coming from another parent and 2 types with a combination of two parents (Figure 5.6). Crossing over, however, results due to the distance of these chromosomes, which would bring about diversity in species.

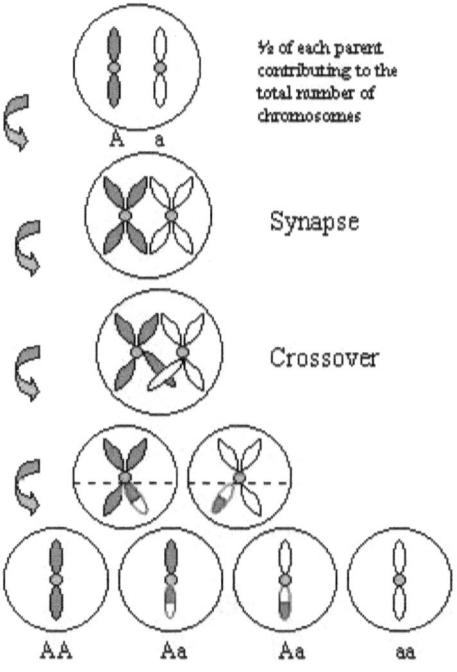

Figure 5.8. Crossover During Meiosis and the Resulting Daughter Cells

The diversity brought about by this process has been the result of random occurrences transpiring at the syngamy or fertilization of two gametes. No one would ever be sure, through scientific principle, which sperm passing half of the total number of chromosomes and bearing distinct hereditary traits will fertilize the egg which also bear half of the total number of chromosomes with particular hereditary properties. Probability could only produce a description of the process. Even then while gametes undergo meiotic division, crossing over and recombination that results in the shuffling of hereditary traits occur at random. Thus with this process characterized with randomness, the diversity of life is produced and life is reproduced in a milieu that is constantly under construction.

Humans as Builders of their Milieu

If we go back to Plato's Allegory of the Cave, we will find that the man who was able to escape from his chains and saw the real objects that cast shadows on the cave's wall discovered experiences for himself. This is a representation of what the milieu-already-constructed presents. But when he went back to the cave and relayed what he discovered to his friend who was then still locked in chains, then what he did was a reconstruction of his experiences to the other guy. He was then building and rebuilding his milieu and the other person's. This is therefore a picture of what the milieu-under-construction represents.

The construction and reconstruction of the milieu proceeds from the nature of humans as *self-inventing* beings. It is their nature aside from their characteristics as being rational and self-interpreting. These two natures of human beings have been the seat of controversial claims between the quantitative and qualitative social sciences. The rational person was first the subject of Descartes's argument in his rationalist school of epistemology that humans are thinking, doubting beings capable of producing knowledge. The mind of a human being is not entirely like a sponge that needs to be soaked to accumulate water or to be immersed in experiences to produce knowledge. Human beings are imbued with the faculty of thought to produce what he knows. This claim is pushed farther by the quantitative social sciences which assumes that human beings, in view of a certain goal would capitalize his gain and minimize his loss. A man who would shop for a shirt would buy the best shirt his money could afford. Either he would save to buy an expensive one or wait for a time to buy the same shirt during a sale if he doesn't have enough money to purchase it. A man who would like to walk towards a certain destination would opt to walk on the road with the shortest distance, unless he re-defines his goal by enjoying a long walk before arriving at his destination. The rational man then is an economic man who is a maximizer of gain and minimizer of cost.

But apart from being rational, human beings are also self-interpreting (Taylor 1985). This is the assertion of the qualitative social science that assumes human beings to be capable of producing and reproducing meanings that are uncovered from observable social action and translated into social practice (Taylor 1979: 49). Peter Berger and Thomas Luckman attest that:

> The world of everyday life presents itself as a reality interpreted by men and subjectively meaningful to them as coherent world... The world of everyday life is not only taken for granted as reality by the ordinary members of society in the subjectively meaningful conduct of their lives. It is a world that originates in their thoughts and actions and is maintained as real by them (Berger and Luckman 1993; 291).

Our everyday life is also a zone of our intersubjective world... "a world that we share with others" (Berger and Luckman 1993: 293). This intersubjective world is a region of the individual's subjectivity meeting at the moment of interaction and creating a network of meanings exchanged and shared through symbols. In his book, *Society as Symbolic Interaction*, Herbert Blumer argues that:

> The term symbolic interaction refers to the particular and distinctive character of interaction as it takes place between human beings. The peculiarity consists in the fact that human beings interpret or "define" each other's actions instead of merely reacting to them. The "response" is not made directly to the response of one another but instead is based on the meaning which they attach to such actions. Thus human action is mediated by the use of symbols, by interpretation, or by ascertaining the meaning of one another's actions. This mediation is equivalent to inserting a process of interpretation between stimulus and response in the case of human behavior (Blumer 1993: 303).

In this context, meaning is being built in the region of interaction. Symbolic interaction assumes that an individual is not "surrounded by an environment of pre-existing objects which play upon him and call forth his behavior" (Blumer 1993: 304) An individual then constructs "objects" as he goes about his activity. The meaning then that individual shares is not simply a "release" or "meaning simply dished out." Symbolic interaction presupposes that:

> Human society is made up of individuals who have selves (that is, make indications to themselves); that individual action is a construction and not a release, being built up by the individual through noting and interpreting features of the situations in which he acts; that group of collective action consists of the aligning of individual actions brought about by the individual's interpreting or taking into account each other's actions (Blumer 1993: 305).

In order to uncover the meaning in the interaction site, symbolic interaction concerns itself with the process as the meaning emerges.

> Insofar as sociologists or students of human society are concerned with the behavior of acting units, the position of symbolic interaction requires the student to catch the process of interpretation through which they construct their actions. This process is not to be caught merely by turning to conditions which are antecedent to the process. Such antecedent conditions are helpful in understanding the process insofar as they enter into it... (Blumer 1993: 308).

Adding to the rational and self-interpreting nature of human beings is his self-inventing characteristic. Since theories have the capability to capture emerging phenomenon, textualizing this occurrence would need closure and either recreate the phenomenon and observe it in a controlled environment or observe it in its natural setting. But human beings are self-inventing agents who are governed by relationships and are driven to create them. Relationships, however, are boundary crossing. Hardly could they be subjected into closure. In order to textualize the patterns of relations in an emerging or on-going phenomenon, it would require that the observer be on the actual site of occurrence, observe what is going on and participate in it. This technique has been the preoccupation of Grounded Theory which the qualitative school of social science has been preoccupied with. Grounded Theory was first espoused by Glaser and Strauss with the publication of their book *The Discovery of Grounded Theory* in 1967. Their work was a revolution against the deductive and positivistic nature of doing science (Glaser and Strauss 1967). A social theorist concerned with the construction of Grounded Theory is interested in the patterns of action, interaction and in the process of mutual exchanges. For this:

> Grounded theory is a set of plausible relationships proposed among concepts and sets of concepts. Grounded theory is always traceable from the data that give rise to them – within the interactive context of data collecting and data analyzing, in which the analyst is also a crucially significant interactant. Likewise, grounded theory is always fluid because it embraces the interaction of multiple actors, and because it emphasizes temporality and process (Strauss and Corbin 1999: 80-81).

In this regard, concepts are formulated and analytically developed, conceptual relationships are posited but they are not constructed out of the voice or perspective of one person, much less the observer, but they are inclusive of the multiple voices and perspectives of the interacting actors. Thus Grounded Theories, which are system-statements just like any other theories are nevertheless grounded directly and indirectly on the perspectives of the diverse action toward the phenomenon (Strauss and Corbin 1999: 84). Grounded Theory, however, is not like the theory that accounts for the milieu-already-constructed, which attempts to produce generalizations about the phenomenon of the same nature. Grounded Theory is inductive. It veers away from the natural science's preoccupation in formulating theories first and testing them in an experiment. The formulation of Grounded Theory emanates from empirical observations and experiences "on the ground." These observations and experiences are conceptualized and rationalized and from which a theory is constructed.

Grounded Theory captures a particular incident in time. It uncovers the uniqueness of the patterns of interactants and relations taking place which may not be true to other phenomenon of the same category or may not be true anymore as time goes on. The patterns though unique are also enduring for they are the result of the enduring processes of interaction. But over time, they could change, thus while enduring, they are also temporal and particular in terms of their location in space and time.

To illustrate this, let's take a traffic situation.

*Textualizing Human Interaction in a Cramp Space**

Traffic is actually a situation that takes place in a certain space where "there is not enough room to maneuver for too many vehicles or pedestrians." The limitation in space creates the production and reproduction of relations and meanings as participants enter into such a limited space. This produces a network of meanings where language, gestures and symbols are traded, exchanged or reinforced. Visit the Philippines, specifically, Metro Manila and you will experience traffic not like anywhere else in the world. For Metro Manila residents, traffic is a usual social gathering where he or she is enmeshed specifically at 7 o' clock in the morning and 6 o'clock in the evening. These peak hours even extend if there is such a heavy volume of vehicles or an accident has happened on the busy street. Traffic is simply a situation in space where individuals participate in the simultaneity and instantaneity of time as they share a common public moment in that certain space by crossing the street, getting off the intersection or wanting to get a ride. Traffic is the sharing of public time in a public space, only that the space is cramp that the participants would have to squeeze in to find some room and arrive at their destination. This cramp space is

---

* This is part of the paper submitted to Dr. Nanette Dungo University of the Philippines in October 2000. Since that time new structures would have been constructed.

The Milieu-under-construction

Metro Manila's everyday life that becomes a melting pot of different actors and objects.

Once these actors interact with each other and with these objects, a web or network of meanings is constructed in turn. The participants form meanings toward each other and towards the objects as they interact in that space. These meanings are not simple definitions but intersubjective meanings or meanings translated into action and social practice (Taylor 1979: 49). These meanings are discernible in the language they use, the actions and gestures they manifest, and the way they use their vehicles specifically through the honks they air. It would be observed that the traffic intersection in Metro Manila is one of the noisiest places, besides it being a honking arena. Since honking is a use of technology and means where one driver communicates with other drivers, different patterns of honks mean different things.

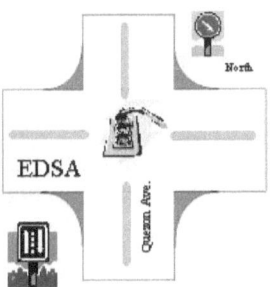

Figure 5.7 The EDSA-Quezon Ave. Intersection

In this regard, I employed various techniques in order to uncover this web of meanings. I used observation where I stood at various corners of the Epifanio Delos Santos Avenue - Quezon Avenue (EDSA-QA) intersection and observed the pattern of occurrences. Being part of the commuting public, I took trips down the intersection aboard a jeepney where I became a participant-observer, sat near the jeepney driver, observed his gestures and listened to his utterances at the moment we encountered other actors and objects in the intersection. It has to be noted that the traffic intersection is a very busy space. You can't interview a driver especially while he is driving unless he opens up and initiates a conversation. It is also difficult to talk to a traffic enforcer while he or she is manning the traffic neither the passengers waiting for a ride. Most often, they would leave you once a jeepney arrives. I embarked on a focus group discussion among groups of Philippine Women's University (PWU-Quezon City Campus) students who frequently crossed and used the intersection. With that I also interviewed drivers plying the Zabarte-Panay, University of the Philippines (UP) -Pantranco routes and street vendors along the intersection.

The EDSA-QA intersection, however, is a busy junction early in the morning and early in the evening. This may even extend late at night depending on the volume of vehicles or if any accident has occurred in the vicinity. It is a high traffic area on all directions since the Southeast direction going Ortigas-Makati-Pasay is a commercial and business district; the Northwest direction towards Balintawak-Monumento-Novaliches is a commercial area (Monumento), residential district (Novaliches), and an outlet (Balintawak) for provincial busses going North and West of Luzon. The direction towards Fairview is both residential and business area where private and government offices, together with schools and universities, are located. The Southwest direction towards Quiapo is both a commercial district and a university belt. Besides being a high traffic area for vehicles, it also accommodates high volume of pedestrians. EDSA is a six-lane highway while Quezon Avenue is a four-lane road. Both streets are separated by center islands on alternate direction added to corner islands on all road interchanges. Provisions for non-ambulant handicaps are in place but some cracks, depressions, small potholes and uneven asphalt patches are existent on the pavement. Besides the traffic lights, several traffic signs, road names and directions are situated in strategic areas though commercial billboards hug the line of sight in the background. The Manila Seedling Bank Foundation covers the Northern section of the corner while on the opposite site in the East is a vacant lot. The Western area is lined with fixed stalls that vend food, while the Southern side is stretched with videoke bars. Elevated on this intersection along the EDSA route stands the stretch of Manila Rail Traffic tracks. This intersection floods during heavy downpour. This research was carried out between June and October 2000. Since then, the landscape could have been altered and new infrastructures could have been built.

But going back to the former landscape, once an individual enters in this intersection, he or she may be considered a participant in the relations evolving in that space. The participants can be categorized into two main types: a) those who are riding in a vehicle and b) those who are on the ground.

a) Those who are riding in a vehicle include the drivers and passengers. Anyone who alights the vehicle may be classified under the second category.

b) Those who are on the ground consist of the pedestrians who walk towards a certain direction or try to catch a ride; the traffic enforcers consisting of Metro Manila Development Authority (MMDA) or Police agents; and the economic takers consisting of food, newspaper, cigarette, candy, face towel, bottled water vendors and street children.

These individuals interact with themselves as they enter into that space. Aside from the people who interact, they also respond to technology and objects either stationary or mobile. The vehicle itself is an object. The honks it issues make other drivers respond and pedestrians wary of its presence. The movement of the vehicle as it accelerates, decelerates, brakes, or swerves are manifestations of the drivers' action and reaction. The whistle, which the MMDA or traffic policeman blows, is an object where individuals in the traffic area also respond to. His motorcycle especially when parked in a strategic area along a "no loading unloading area" creates an image that the traffic agent is just within the vicinity. The painted lines on the pavement are controlling objects which both drivers and pedestrians are constrained where to pass and enabled where to move along. But these lines are not clearly defined on the pavement specifically on this intersection. The traffic lights on all directions are prominent controlling technology but this automatic device can be manipulated manually. A control box is situated at the Western corner of the intersection which the traffic agents could open, call the central station and manipulate the lights at their choosing if the unusual heavy volume of vehicles occur on either direction. Traffic signs are well placed at strategic points.

We need to understand that a traffic intersection is a busy area. People there are in a hurry. And because it is an intersection where vehicles and pedestrians intersect their paths, it has to be controlled. The control is realized using traffic lights, traffic signs, and the traffic enforcer issuing signs and whistles. With the data gathered, the traffic intersection yields some interesting findings:

The traffic intersection is a danger zone. Pedestrians hurry up as they cross because they are afraid (*natatakot*) for they might get hit by on-rushing vehicles. They don't regard the traffic light when they cross because they presume it is not being followed anyway; it is defective and sometimes it is turned off. The traffic light is simply an ornament. Pedestrians cross when oncoming vehicles have stopped, or when there are a few vehicles passing through. At this moment, it is already presumed to be safe. The signal to cross is not the light but other pedestrians (*ang iba*) who have successfully crossed or when everybody is already crossing the lane. Instead of the intersection being predictable because of the traceable pattern of green and red lights at certain intervals, it is rather a hazardous, unpredictable site (*nakakalito*) where traffic enforcers can arbitrarily alter the traffic lights' patterns, even disregard them when the traffic enforcers are the ones crossing. Instead of the intersection being a controlled area using the traffic lights, the gadgets are regarded as simple ornaments on the street. The traffic light, though it serves its purpose of putting vehicles to a stop or go mode, is perceived as means of entrapment. Since policemen can arbitrarily manipulate the light, they can quickly alter green to red and catch incidental violators.

- Pedestrians and drivers define traffic laws not by the sign posted on the corner, but on the location where the traffic enforcers are situated. The traffic law (*bawal*) is that certain space behind the traffic enforcer away from his view. Drivers stop at that location behind the traffic enforcer away from his sight without regard to where the traffic sign is located. Commuters expect the same, for they all define a traffic enforcer to be corrupt (*buwaya*) and abusive of authority, lacking in credibility (*walang alam tingnan*) who simply apprehends (*nanghuhuli*) rather than maintains order. The space in front and beside a stationary traffic enforcer on the look-out, or his parked motorcycle is a "no loading and unloading area" even if a sign to the contrary is standing on the corner. It is a forbidden zone. Pedestrians and drivers recognize the presence of these traffic agents because they are afraid of getting caught. They fear apprehension because they don't want to be fined, since it is presumed that traffic agents are corrupt. To get caught is to be fined (*huli*). Since traffic laws are defined by that space where traffic agents stand, without these enforcers, there is no law (*hindi bawal*) and vehicles could just stop, load and unload anywhere except in the middle of the intersection which is a high danger zone and the possibility of getting rammed is high. For pedestrians, they would rather cross, alight, or get a ride anywhere for convenience. If traffic agents are presumed to be corrupt (*buwaya*), drivers, among themselves, also admit that without traffic enforcers, they are also abusive (*barubal*). While traffic agents are perceived to be corrupt, pedestrians also admit of violating certain rules, thus they run while crossing so as not to get caught.

- Traffic is a zone where honks mean different things. Sets of two or three staccato or detached short honks, coupled with the driver or conductor waving his hand with palms down and shouting his destination means "there are still some seats or space inside the bus or jeepney and we could take you to a certain destination." Once vehicles cross and begin accelerating through the intersection, sets of two or three longer than the staccato honks mean, "hurry up, speed up." If the vehicles accelerate by overtaking a vehicle that is about to swerve, two or three longer honks mean "I am here you might hit me." If the vehicles start moving while a vehicle suddenly get stalled or has not moved, one long honk coupled with short ones, mean, "you are blocking my path." If a "long loud piercing honk with the word uttered "*Gago* (stupid)" to the driver of the vehicle in front means that the other driver is a nuisance in his path, a stupid driver who does not know how to drive as well as he could. These patterns of honks are not learned in a driving school. They automatically emerge once a participant gets immersed in such a situation.

- The honks testify of a hazardous, busy space which needs to be maximized. There is not enough room

The Milieu-under-construction

to maneuver in traffic. Thus a skillful driver is presumed to be one who could take advantage of a cramp space. A skillful driver is one who can find alternate route in a massive build-up of vehicles or one who could maximize space from one point to the other at the least possible time, even if it means squeezing the lane and passing other vehicles (*singit*) and even if it means violating an institutionalized traffic rule (*lusot*) as long as he does not get caught, hurt someone; sideswipe or hit another vehicle. A car crash is a violation of someone else's space. A driver honks because he needs to occupy the space which the vehicle in front occupies. The vehicle will also be honked if he tries to swerve as a warning that he would be occupying the space he would rather try to occupy. The crash happens because the vehicle suddenly tries to occupy the space the others occupy, displacing, sideswiping or bumping the other vehicle. Thus a good driver can pass by all these with his vehicle free of dent or scratch. The same is true with pedestrians who hurry up crossing the street. They are afraid of being caught and are taking chances at crossing it successfully (*nakalusot*). Drivers obey the law for fear of apprehension and fear of paying penalties.

- In this cramp space, economic opportunists consider this as means of livelihood. The ambulant food, newspaper, bottled water, rags, face towel vendors and street children consider this cramp space with vehicles and people as an opportunity to earn money and a chance to get out of their life's misery. Moreover, for the street children, the cramp space is not simply an area open for economic opportunity but a playground. While they earn money, they also have fun.

The conspicuous device in a traffic intersection is the traffic light. Drivers respond to it. Red means stop. Green means go. But a concoction of other responses in regard to the traffic light creates other intersubjective meanings towards the device. Yellow by common definition means slow down. But drivers all the more accelerate and honk loudly to tell the intersecting vehicles, "Wait, I'm still here, I'll pass." At the green light, while vehicles rush to move across the intersection, sets of two or three long honks are aired to tell others to speed up. This honking and space scampering stop once the red light lights up. Once the light turns green, those who are slow to move will receive honks which mean, "Move faster, you're blocking my path."

But for pedestrians and economic takers, the traffic light is not the devise that signals the moment to cross or to resume their work on the street. Movement is the signal. As soon as vehicles stop, pedestrians cross provided they could sense it is already safe. The movement of other pedestrians as they are about to cross or have successfully crossed also present themselves as signal for others to cross the street. The honk is issued by virtue of the movement discernible at the nose of the vehicle which tries to move left or right.

At this point, we could separate institutionalized from intersubjective rules. Institutionalized rules are legislated laws or ordinances promulgated by the state. The intersubjective rules are norms born out of intersubjective meanings discernible in action and social practice. Traffic signs as operationalized traffic rules are strategically located but there is a different definition of these rules once socially practiced. Any traffic rule is defined by space. The space where the traffic agent is located defines what the traffic law is all about. The space behind the traffic agent is a permissible area, whether to load or unload passengers while in front of him is a prohibited space, disregarding whatever the traffic sign is posted on the wayside. But it is not simply the location of the traffic agent. The gaze and position are important. What defines a traffic agent to be imbued with a role to apprehend, as perceived by every participant in the space is the stance:

arms on the side, waving towards on-coming vehicles to compel them to move on quickly, sometimes blowing his whistle if others are not quick to respond. The gaze too is important. The sharp eyes focused far symbolize a traffic agent in action. Even then without the traffic agents, the mere appearance of a motorcycle in the area is a cause for concern since the gaze could be anywhere hiding which could be a means of entrapment. But drivers and pedestrians know the permissible area for they will load and unload beyond the parked motorcycle.

The space also should be maximized. The maximization defines a skillful driver. It is even tolerable and permissible for traffic enforcers that vehicles start to occupy the middle of the intersection for those who will turn left, even if the green light for "left turn" is still off.

In this research, we find the concept of space and the web of meanings being constructed in it. The Grounded Theory then would proceed in this manner:

> In traffic there is not enough space to maneuver for too many vehicles. While it is intended to be a controlled area where traffic lights are a technological means to maintain an orderly flow of vehicles and pedestrians, it is perceived to be a dangerous (nakakatakot) and unpredictable space (nakakalito). It is a busy space defined by the honks they issue in order to maximize the space. In its crampness, it creates a situation where the traffic law is defined by the presence of a traffic enforcer on the lookout or just his motorcycle stationed in a certain space. Where the traffic enforcer stands or motorcycle parked defines the law; behind him or her is the space free of apprehension (huli) and beside or in front of the traffic enforcer is a forbidden area. Without the traffic enforcer there is no law. In the same manner the space where he stands defines him or her to be corrupt (buwaya) and abusive of authority and whose work is defined by merely apprehending violators (nanghuhuli) and not keeping order. In that limited space

what defines a skillful driver is one who could maximize space from one point to the other at the least possible time by squeezing and swerving in lanes and passing others (singit), even if the act violates institutionalized rules. But he could get away with it, by not getting caught and not hurting or being hit by others (lusot). The violation of space, however, defines a car crash. One vehicle wants to occupy the space others occupy. Traffic is also a space of opportunity to find livelihood and means to get out of life's misery.

Traffic is a space where relations and interactions exist. Out of this interaction in a cramp space, meanings evolve. But these meanings do not only come about as a way of being released or dished out. It is not even simply a confluence of meanings. Rather, it is a "feeding" of meanings. The participants feed on each others' meanings, thus a match of meanings is secured. The mutual feeding creates a web or network of meanings constructed in the cramp, limited space. This web of meanings is the answer why we find a seeming order despite the chaotic location. One can observe the breakdown of institutionalized rules (chaos) in this situation. Passenger jeepneys loading and unloading anywhere even if "no loading and unloading" signs are fixed. Pedestrians cross without regard to traffic lights. Traffic lights are sometimes on or sometimes off. There is the unending honk, there is the pervasive swerving, cutting, and overtaking just to maximize space. This defines chaos where institutionalized rules are blatantly violated. But once a participant enters this cramp space and interacts with the actors or objects in it, the chaotic space becomes coherent.

Passengers load and unload anywhere without regard to traffic signs because the law is defined by the space where traffic personnel is located. Without the traffic agent or his equipment, any space as long as it would not cause harm to others become a permissible area to load and unload passengers. Passengers also ask to board off (para) and get a ride at such places without the presence of the traffic personnel. In the same manner, the only job they could think of these traffic agents is not to keep order but to apprehend violators. Passengers and drivers evade their gaze and their presence because they don't want to get caught and be penalized monetarily. Getting caught is getting fined. Pedestrians cross without regard to traffic lights because the light is not the signal for them to cross. The signal emanates from the movement of the vehicles, as soon as the vehicles have stopped, or the movement of fellow pedestrians as they are already crossing or have successfully crossed the street. The honk does not simply mean "beware" but it ranges from "There is still space inside my jeepney" to "Get out of my path you fool, you're blocking my path." Traffic enforcers, either remedy an unusual volume of vehicles or trap and catch unwary offenders, by arbitrarily manipulating the traffic lights. This is a space in which swerving and squeezing lanes is done to maximize space. This is the mutual feeding of meanings. This is the match of meanings. Here we find order amid chaos. One can understand it if he or she enters that limited space, becomes a participant of the mutual feeding of meanings and becomes enmeshed in the web of meanings created. Thus the concept of space is interactively and intersubjectively constructed. But this concept of space is reinforced by the concept of movement (as signal) and the concept of gaze (as authority).

The web of meanings does not only explain the intersubjective order amid chaos in terms of the breakdown of institutionalized rules but also explains why at some areas these same drivers would behave differently once they get out of this web. The researcher observed that these drivers (specifically UP-Pantranco) obey traffic signs and institutionalized rules even without the presence of policemen, once they enter the University of the Philippines (UP) Campus. The moment they get out of UP Campus, they become blatant violators of institutionalized rules which they follow inside the university. Here is the concept of space at work.

The University of the Philippines is a different space altogether. In this space a different web of meanings is created. Thus a different set of actions and reactions come about. A different set of behavior is manifested. But what reinforces the difference? The answer is on account of the sense of community. Those who live and identify themselves with the University of the Philippines or any other organization or geographical location develop a sense of community. This again is space where a sense of community emerges. Drivers plying the route inside and outside UP have an organization depending on their destinations. They are required to be members of the driver's association and be issued with identification cards authorized by the UP administration. The ID, the route, their personal knowledge of their co-participants in that space, their common understanding of that space develop in them a sense of community in which they define the rules complementary to institutionalized ones. Thus the stronger the sense of community in that certain space, the more defined the rules, intersubjectively and institutionally.

Who would identify himself with a traffic intersection? The site is a host of strangers where even one doesn't know who his neighbor is in a city. The concept of space in the city is not the whole city but limited to a few neighbors, a compound, a workplace or even the floor and roof of a house. This is how limited the concept of space is in a city, which unlike in a village, one knows of someone even in far-flung areas and identifies himself in that community. Without the sense of community in a public space trust is lost. Thus if "everybody is violating the traffic law why can't I." The presumption is that everybody is violating the institutionalized rules anyway. The mutual mistrust feeds on the web of meanings creating a different intersubjective rule apart from institutionalized ordinances. An intersection is a public space, where every participant interacts with each other and with objects at a common public time. No one lives permanently in an intersection where a sense of community emerges. It is a public space, a public space of strangers, a public space defined by movement and gaze. But in this intersection, meanings and social practices are built and reinforced and the milieu of the participants is created and recreated.

Humans as Self-Inventing Agents

What sets the social sciences different from the physical sciences is the nature of the human agent which the social sciences examine. A geologist who studies rocks examines inanimate objects whose nature, structure and dynamics are highly dependent on their surroundings. A quantum physicist investigates the behavior of electrons, calculates the position, movement and the force that make the particle move. But neither the rocks nor the electrons have minds of their own to alter their movement, change their structure or modify their environment at will. On the other hand, the rational, self-interpreting human being has the capability to invent himself even intervene in his environment at will as motivated by his intention or compelled by the forces emergent in his environment in the direction of his goal. He is not like the planet which orbits around a certain body in space in its lifetime unless an outside force directs its movement some place else. A human being, on the other hand, could decide on the path he would take, either take the mainstream or tread the way of his own.

Thus besides the constitution of the human being as imbued with rationality and self-interpretation, he is also an autonomous agent capable of making decisions, though the decisions he dispenses may have been fashioned out of the concoction of several internal or external influences. This is matched by the constitution of human agent to "will." Human will provides the autonomous agent with the capability to decide and carry out such decisions into social action. These constitutions present the human agent as self-inventing being who does not only intervene in the environment where he is constituted but also invents himself to adapt in his environment. This makes the human agent unique. And it is this very reason why the social sciences have a difficulty formulating generalizations and generating law-like assumptions unlike the physical sciences. Take the Pavlovian experiment for example. The Russian Noble Prize winner was actually working on the study of digestion when he later theorized on classical conditioning.

Ivan Pavlov subjected a dog to an experiment where at some precise interval, he would introduce food to the dog at the switching on of the light.

> A dog is prepared for Pavlov's experiment by having a minor operation performed on its cheek so that part of the salivary gland is exposed to the surface. A capsule is attached to the cheek to measure salivary flow. The dog is brought to a soundproof laboratory on several occasions and is placed in a harness on a table. The preliminary training is needed so the animal will stand quietly in the harness once the experiment begins. The laboratory is so arranged that meat powder can be delivered to a pan in front of the dog by remote control. Salivation is recorded automatically. The experimenter can view the animal through a one-way glass panel, but the dog is alone in the laboratory, isolated from extraneous sights and noises. A light is turned on. The dog may move a bit, but does not salivate. After a few seconds, meat powder is delivered; the dog is hungry and eats. The recording device registers copious salivation. The procedure is repeated a number of times. Then the experimenter turns on the light but does not deliver any meat power. The dog

salivates nonetheless. It has learned to associate the light with food (Atkinson et. al 1983: 194).

This is classical conditioning at work which became one of the theoretical foundations of learning applied to human beings. While the basic assumption that human beings learn by association, the idea of learning may also make him deviate from the direction which the assumptions would lead him.

A child who gets spanked for spilling milk may associate the spanking with being careful when handling the glass of milk, lest she get smacked again. But even if she is able to learn painfully along the process, she may still choose to spill the milk to test the resolve of her Mom, spill it intentionally and clean it up stealthily, or point to others as the culprit to escape the punishment. Thus with respect to behavioral sciences, humans can hardly be subjected to experiments and still come up with accurate and objective findings with high degree of predictability and replicability.

Even survey results that underwent statistical treatment to determine cause or effect would generate problems. If it were found out in a survey that strong affiliations with homosexual friends, a domineering mother and weak father figure would lead to a high tendency of homosexuality, a young person who finds himself in the same situation as revealed in the statistical results, may either box himself in to justify his becoming a homosexual later or may re-invent himself to make a stand that such findings don't apply to him. He may have homosexual friends, a domineering mother, a weak father but he is not gay. Here even the survey results, treated with statistical procedures could influence the individual to make him re-invent himself to conform with the findings or to go against them.

If we go back to the Pavlovian experiment and place a human being as a the object of conditioning under the same experimental set-up instead of a dog, we may find the humans behaving quite differently for he might get pissed off at the turning on or off of the light though he may actually associate the light with his meal. But it would also be a possibility that upon learning it, he may choose to refrain from eating his meal at the switching on of the light or eat it afterwards, ruining the predictability of the experiment and destroying the findings altogether. Had this occurred, then the erratic results would then affect the researcher which would force him to create a new experimental set-up. With humans as experimented subjects and humans also doing the experiment, we find both of them influencing each other. And inasmuch as the experimented subject is influenced by the instruments and the procedure of the experiment, the readers and consumers, who are humans themselves, of the findings will also be influenced by the results.

This illustrates the nature of humans as self-inventing beings and their actions as recursive-interactive in essence. If we move out of the laboratory and capture this phenomenon in its natural setting, we could find the library as an example. The library, however, is presumed to be a quiet area. Let's say two library users were conversing loud enough that their voices were disturbing other library patrons. In order to suppress the noise, the librarian decided to stand up from the library counter and with a stern look shoos up the two individuals, loud enough for the two library users to be warned but also loud enough for other library patrons to be disturbed. The point is, after the brief confrontation with the shooing up as the language of communication, the other library patrons around them may have also been disturbed that they would end up staring at the two noisy individuals who, by then, had already folded up their lips at the sight of the pissed off librarian.

The Milieu-under-construction

In this scenario, we find that the action issued by the librarian was brought about by the antecedent cause created by the noisy library users. Her decision was influenced by the purpose of the library to create a quiet environment conducive to learning but as she acted on those influences she disposed of an action that influenced the noisy library users and the quiet library patrons around them. The reaction of the two individuals as they went back to their reading and the other patrons staring at them for creating an equally noisy expression fed back on her that in turn made her realize that she, herself, also made a noise. In that instance, the condition in the library as an environmental influence was altered. This illustrates the recursive-interactive nature of an act that makes for a self-inventing being. The action of an autonomous, rational, self-interpreting being, which was formulated out of several influences, affects the object of his act and affects him as well which presents him for possible re-invention of himself or possible modification of the influences. The human agent's act re-invents the object of his act which in turn re-invents him as well and constructs or reconstructs the environment or condition from where the act takes place. The continuous motion of this schema allows the individual to build and rebuild his milieu, making for a milieu-under-construction.

If we go back to the experiment, the environment in the experiment interacts with the experimented subject and gives erratic results. At that time, the experimented subject re-creates himself and provides results which would also re-invent the researcher. In order to rectify the experiment, the researcher would change the set-up, re-creating the environment where the experiment takes place. Survey results also create a similar effect. If it were shown that 75% of those who abuse their wives were also abused children before, then an individual who does not abuse his wife may associate himself with the 25% who did not suffer child abuse when they were young. On the other hand, wife beaters who never had any remembrance of their childhood may assert to themselves that the reason why they abuse their wives is because of the child abuse they suffered although they may uniquely have other factors to influence them individually. In the same case, a person who knows he suffered from child abuse now realizes that he would have a tendency to perpetuate domestic violence and tries to check himself in order not to harm his wife. Just the same, the survey results have re-created the person and another survey may not confirm the same findings if most of the 75% would change their violent patterns. In the same way, theories that textualize the milieu also re-create human perception and action to comply within the patterns set by the theory. Take Karl Marx's theory for example.

Textualizing Economic and Political Relations

In the language of theory, the specific concern which has shaped Karl Marx's theory is the exploitative condition of the working class in view of the economic relations that they share. Remember the onions that we were trying to figure out in the market. Let's say it is produced by a plantation of onions owned by an agricultural capitalist. The production process, from the planting, harvesting, storage, packaging and distribution of onions, is owned by the capitalist but dispensed with by the agricultural workers. Since the profit derived from the sale of the onions reverts to the capitalist, the workers earn their compensation from the wages given by the capitalist. With this economic relation, what would happen if the workers try to ask for higher wage? In the language of exploitation, Marx assumes that the owner of the means of production dictates the political and legal superstructure of society.

The Milieu-under-construction

> In the social production which men carry on they enter into definite relations that are indispensable and independent of their will; these relations of production correspond to a definite stage of development of their material powers of production. The sum total of these relations of production constitutes the economic structure of society – the real foundation on which rise the legal and political superstructures and to which correspond definite forms of social consciousness (Marx 1981: 174).

The productive relations end up towards the domination of the class which owns the productive means of production from which the political structure of the society is built. As an economic determinist, Marx fused his ideas with the critique of his fellow thinkers to come up with his own Dialectic-Historical-Materialism. It is his theory of society that subscribes to Georg Frederick Hegel's dialectics which became the foundation of his idea. Hegel assumed that:

> Every condition of thought or of things - every idea and every situation in the world leads irresistibly to its opposite and then unites with to form a higher or more complex whole. For not only do thoughts develop and evolve according to this "dialectic movement" but things as equally; every condition of affairs contains a contradiction which evolution must resolve by a reconciling unity. (Durant 1961: 295-296).

The idea of dialectics brings to fore the emergence of thesis-antithesis-synthesis mode of analysis. This means that a condition (thesis) creates its own contradiction (anti-thesis) which will bring about the emergence of a new condition (synthesis). Suppose the government passes a state legislation putting a cap on salary increases. This condition would breed its opposition from the labor sector. In the same manner, a state legislation raising the salary of workers would encourage protest from the businessmen for it would defeat their freedom to adjust the salary of their employees commensurate to the earnings of the company. If these contradicting situations persist then a new condition could come up which may be a modification of the initial condition or an entirely different situation may emerge out of it. Inasmuch as Hegel believed history proceeds in this manner, Marx subscribed to the same idea but rejects idealism which Hegel espoused. Hegel believed in the importance of the mind and mental products rather than in the material world. Ideas to him are real and the material world is simply a product of these ideas. Ludwig Feurbach challenged this idea and advanced materialism in return. Feurbach argues that matter is real and that human beings are a material reality (Ritzer 1996: 22-23). But Feurbach pushed his ideas on the realm of spirituality, postulating that God has become "a projection by people of their human essence onto an impersonal force where people set God over and above themselves with the result that they become alienated from God and project a sense of positive characteristics onto God while they reduce themselves to being imperfect, powerless and sinful" (Ritzer 1996: 23). Feurbach added that this thinking could be overcome by materialism where religion as an abstract idea becomes subordinate to the material man being the highest object.

While Marx subscribed to Feurbach's critique, Marx, however, did not agree to his preoccupation on religion and pushed his idea that economics and productive relations are the frontiers which materialism could theorize. He argued:

> The mode of production of material life conditions the social, political and intellectual life process in general. It is not the consciousness of men that determines their beings, but on the contrary, their social being that determines their consciousness (Marx 198: 174).

The Milieu-under-construction

With this, Marx built his work on the ideas of other political economists such as Adam Smith and David Ricardo who advanced the primacy of labor as the true measure of wage. If labor becomes the true measure of one's emolument, then Marx challenged that in productive relations, why is it that the owner of capital is the one who gets the most of the surplus and not the laborer? Here Marx weaved in history with dialectics and materialism. Marx believed that in view of economic and productive relations, society proceeds from feudalism to capitalism as interspersed with dialectical conditions.

In a feudal society productive means are founded on the relationship of humans with the land. It brings about the occurrence of two classes: the landlord and the vassals. In the productive processes, the surplus derived from tilling the land does not go to the vassal but to the landlord who owns the productive means.

> At a certain stage in the development of these means of production and exchange, the conditions under which feudal society produced and exchanged, the feudal organization of agriculture and manufacturing industry became no longer compatible with the already productive forms (Marx and Engels 1964: 66).

As societies develop into capitalism, new productive relations emerge based on the objective to produce manufactured goods and new class stratification results between the owners of the means of production - the capitalist or bourgeoisie on the one hand and the laborers or proletariat on the other (Figure 5.8). The dialectical relations of the two classes emerge from the diversion of the surplus of the productive means to the capitalist. This led Marx to plot his Labor Theory of Value where he argued that given a certain amount of wage for the laborer, the capitalist would require the laborer to increase his productive capacity in order to boost his profit. For the agricultural capitalist who owns the onion plantation, if a worker could package 50 sacks of onions for 8 hours, he would require each worker to increase his quota by 5 more sacks for the same hours even without an increase in emolument. The capitalist did this to increase production whose profit ultimately goes back to him.

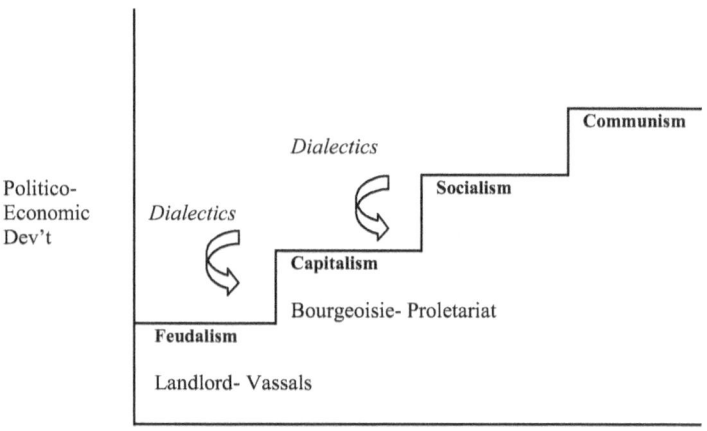

Figure 5.8 Dialectic-Historical-Materialism in Line Plot

Owing to the extensive use of machinery and to the division of labor, the work of the proletariat has lost all individual character and, consequently, all charm for the workman. He becomes an appendage of the machine... Hence, the cost of production of a workman is restricted... to the means of subsistence that he requires for his maintenance, and for the propagation of his race. But the price of a commodity, and... also of labor is equal to its cost of production. In proportion, therefore, as the repulsiveness of the work increases, the wage decreases. What is more, in proportion as the use of machinery and division of labor increases... burden of toil also increases, whether by prolongation of the working hours, by increase of the work exacted in a given time or by increased speed of the machinery (Marx and Engels 1964: 69).

If we now answer the question posed earlier, "what if the farm workers in an onion plantation opt for higher wage, would the capitalist give in to their demand," the theory would predict "no" and the conflict between the two classes would be severe. As the contradiction heightens, the working class would have no recourse but to change the oppressive system, transforming capital from private to state ownership and abolishing class that has brought about exploitation and antagonism. From here he argued that capital is a collective product of the collective endeavor of the members of the society and that "capital is not a personal, but a social power" (Marx and Engels 1964: 83). Thus, with Marx's desire to end the exploitation of men by men, he advanced the change of the system through the change in the ownership of the means of production. If capital is in the hands of the workers having translated it into state ownership, workers then could establish a superstructure free of domination since the oppressive, dominant class has been abolished.

Marx's wish was not only to interpret the world but also to change it. In this way, Marx did not only provide an analytical tool to look at the milieu but also a way to reconstruct it. But how does this theory provide the ingredient for the construction of theory with the objective of changing societies? At this point, we have to bear in mind that the exploitation of the working class is the theme on which Marx's theory is founded. Dialectic-historical-materialism presents a way on how to analyze productive relations that breed class antagonism with the end of eliminating the bourgeoisie in favor of a classless society. But like other theories, Marx's ideas strike at the core of humans as self-inventing individuals. Moreover, it touches on the theme that wealth derived from the exchange of commodities or services, after subtracting the cost of production, is scarce. Profit is scarce. The income of a company is also scarce. And scarce as it may be, the capitalists and the workers are too eager to have a portion of it. The workers, however, who used their skills and bodies to manufacture the commodities and generate the services, outnumber the capitalists whose money was employed to finance the cost of production. Yet at the end of the day, the capitalists get the large portion of the income of the company, if not all of it. A worker could identify himself with the reality of such relations. According to Marx, as the contradiction between the two classes heightens due to increasing exploitation, the workers would then unite and seize the political and legal infrastructure of the society. Marx's works are prescriptive and a little prophetic that the excesses of capitalism would lead to its eventual overthrow by the working class who will erect a socialist system toward a utopian communist society where contradiction would then cease.

> In short, the communists everywhere support every revolutionary movement against the existing social and political order of things... They openly declare that their ends can be attained only by the forcible overthrow of all existing social conditions. Let the ruling classes tremble at a communistic revolution. The Proletarians have nothing to live but their chains... Working men of all countries, unite! (Marx and Engels 1964: 116)

An aggrieved worker as a self-inventing individual would recreate himself according to Marx's assumptions and prescriptions in order to replace the social infrastructure. But not all individuals would subscribe to the same prescription. One criticism thrown at Marx is his overly linear view of society's history that it progresses from feudalism, capitalism, socialism to communism. But the self-inventing individuals may opt to restructure their system and go back to having a liberal-capitalist economy. The self-inventing individuals could either tailor their thoughts to adapt to the Marxist mode, go against it that the theory collapses on them or even modify the theory itself. The critical school of Herbert Marcuse, Theodore Adorno and Max Horkheimer led in the critique of Marxism in regard to its economic deterministic nature. Others desired to go back to the roots of Marxism and were later called Hegelian Marxists (Ritzer 1996: 142-147). Still others wished to be classified as intellectual Marxists, ethical Marxists or revolutionary Marxists. If there is an example in history how this theory figured out on the self-inventing mold of individuals, the Soviet people are an example.

The big problem that the Russian revolutionaries faced in early 1900 before the birth of the Red Bear, was how to fit in the Marxist theory with the society they were confronted with. Marx was talking of a well-developed capitalist society with intense contradiction between the laborers and the capitalists. But what they had was a backward Russian society with feudal economy and a czarist, monarchical order. They were at a debate whether to short circuit the social process by leaping into a communist society at once or letting the evolutionary process move toward socialism and communism. Even Marx's theory itself had to figure out certain revisions, thus the emergence of Vladimir Lenin who criticized the theory. He argued that by themselves, the workers would not be able to ascend to the communist ideal for they could easily collaborate with the ruling class. He propounded the organization of professional revolutionaries whose job was to advance the revolutionary process. Here we find the self-inventing Russians trying to fit in a theory or trying to modify the theory to fit in to them.

Textualizing the Market

Romanticized by Marx's doctrine, the Soviet revolutionaries steered their country towards a communist economy and took away competition which was stamped as evil. Competition is the capitalist's main cloak and, for the communists, any garb worn by the capitalist had to be stripped off. In a communist economy where capital is state-owned, the state plans the economic targets, prescribing how much the state-owned factories would produce in conformity with the projected needs of the people, even the quality, style and design of the product they will produce. But the evil, which the command economy hoped to wipe out, gave birth to another form of evil. The system defeated individual preferences and outrightly murdered the market. Since the inception of the command economy by the Soviets in early 1900, it eventually died in 1989, permitting the tenets of the free-market to re-emerge.

There are actually various ways of looking at the market. It can be viewed as a system of production and consumption; of capital being translated into goods and services, which are then purchased and consumed by consumers whose income derived from their labor, would be used to buy the goods and out of the sale of the goods, the income goes back to the manufacturer-capitalist in the form of profit. The capitalist would again procure the raw materials, pay the rent and other cost of production in order to produce the same goods and services. Through this system of production and consumption, scarce resources are allocated based on the preferences of each participant in the market who is imbued with freedom of choice and rationality to select which goods or services to procure. This is matched by the

The Milieu-under-construction

efficiency of producers who will manufacture the commodities and services in order to satisfy their needs and preferences. In this view, the market works through the satisfaction of the consumers who become the gauge for producers to generate more or innovate the products. In this case, we have to look beyond the demand curve which is the significant element of the market.

Another way of looking at the market is to view it as an arena of strategy. Consumers and producers are always engaged in making decisions to buy and sell. When, where, what and how are crucial to both of them for it would determine their success in maximizing their resources. This means that the producer and consumer have to decide and act on that decision in order for goods and services to be exchanged most effectively and for money to circulate quickly. If they won't do so, the market will collapse for money, which is also scarce, will not change hands and no one benefits. Both the producer and consumer therefore are engaged in strategies. In order to model this behavior, economists have likened these characteristics of the market into a game; thus emerged the Game Theory.

As the name suggests, Game Theory hopes to simulate real life situations involving pay-off structure with the assumption that players maximize resources available at their disposal. For example, John and Peter are watermelon producers. At peak season, both of them were able to harvest quite a large volume of watermelons which were ready to be sold in the market. But knowing that if the market is flooded with watermelons, the price of the cash crop would drop, taking away part of their profit. Being rational calculators then, they have to choose the best course of action to take. In order to analyze this situation, economists use a matrix (Figure 5.9) to construct the pay-off structure.

	John sell	John not sell
Peter Sell	$3, $3	$5, 0
Peter not sell	0, $5	0, 0

Figure 5.9 Pay-off Structure and Strategy for Peter and John

In this simplified matrix, we find John and Peter to be locked in a decision either to sell or not to sell their watermelons (Figure 5.9). Not selling means delaying the trade of the produce in order to wait for the time the supply of watermelons runs out and a better profit results. But delaying the trade further than expected would age the watermelons and their freshness will be lost. Thus if John chooses not to sell but Peter sells his watermelons, Peter gains $5 profit. The reverse happens if Peter does not sell while John sells his merchandise. If both of them will not sell, no one gains anything but if both of them sell, each will gain some earnings only at a reduced amount of $3 for quite a large amount of watermelons would enter the market. If both farmers know the pay-off structure, then the best strategy to take would be for each of them to sell their watermelons. Both of them will opt to sell, because they think they will substantially gain. Both options now converge to the first cell, making it the optimal choice or the equilibrium.

There is also another type of "deadly" game called the zero-sum game. Unlike the previous example where at a certain point both farmers could go home at the end of the day taking with them certain amount of income though it may fall below their expectation, a zero-sum game is a winner-take-all situation.

The Milieu-under-construction

		Mary	
		buy	not buy
Tess	buy	0, 0	1, 0
	not buy	0, 1	0, 0

Figure 5.10 Zero-Sum Matrix for Tess and Mary

Let's say, at the last day of the bargain season, Mary and Tess (Figure 5.10) with only a limited budget, found themselves at odds to buy the last remaining blouse hanging on a display rack of a flea market. If both will decide not to buy the blouse, naturally they will go home with nothing. If Tess buys it and Mary doesn't, Tess will bring the blouse home. The reverse happens if it were Mary who bought the blouse and Tess didn't. If both of them will go ahead and buy the blouse, the saleslady would have no choice but to give the blouse to someone else at a price higher than what they expect in order to break the seeming odds between them. Both of them would still go home empty-handed. Since each will think, it is beneficial for them to buy the dress ahead of the other, they may still end up buying at the same time and still go home with nothing.

A more complex game, yet an interesting one is the Prisoner's Dilemma. This is a classic game where two suspected criminals were captured and jailed in a separate cell. Both were separately interrogated with the investigator trying to figure out who did the crime. The investigator does not only ask who among the two suspects were guilty but in order to psych up the suspects, the investigator coaxes each one to implicate that it was his companion who did the crime, dangling the incentive that if he does confess that the other fellow was guilty, he will go scot-free. In this condition, the pay-off structure would mean the reduction of one's losses.

		Suspect A	
		Confess "he did it"	Deny "he didn't do it"
Suspect B	Confess "he did it"	-3, -3	0, -5
	Deny "he didn't do it"	-5, 0	-1, -1

Figure 5.11 Prisoner's Dilemma Payoff Structure

    Since each of the prisoners does not know what the other would say while only knowing that by confessing, he would gain back his freedom, the suspect would likely choose to confess (Figure 5.11). If suspect A confesses that his companion did the crime while suspect B denies that Suspect A did it, then Suspect B gets incarcerated for 5 years, that is 5 years less than his liberty. The reverse happens if Suspect A denies that Suspect B did the crime while Suspect B confesses it was Suspect A who did it. If both of them deny that each of them did not do it, still they will need to be incarcerated for one year within the length of the investigation. But if both of them confess, then they will both receive the penalty of 3 years imprisonment only at a reduced sentence for being helpful to implicate each other. Since each suspect does not have any knowledge what decision the other fellow would take, then they are locked in a dilemma and since the best option is to implicate the other, then by doing so, both of them will still languish in jail.

    If we apply this game in an economic situation, we could modify the theoretical underpinnings and fit in the scenario of the two salesmen who are trying to convince a millionaire to buy their insurance policies. Neither salesman knows what the other will tell the millionaire in trying to sell his policy but only the client has contact to the two competing salesmen, asking them what each salesman thinks of the other's policy.

		Salesman A	
		Admit "the other policy is good"	Deny "the other policy is bad"
Salesman B	Admit "the other policy is good"	$0.5, $0.5 million	0, $1 million
	Deny "the other policy is bad"	$1, 0 million	0, 0

Figure 5.12 Application of Prisoner's Dilemma for Salesmen B and A

If both salesmen (Figure 5.12) would say the other insurance policy promises well, the millionaire will only buy half of each policy since his desire is to have only a million worth of insurance. If Salesman A puts down Salesman B's insurance while Salesman B admits Salesman A has a good policy, then the millionaire takes their word for it and takes a million worth of insurance from Salesman A. The reverse happens if Salesman A says Salesman B's insurance does well while Salesman B undermines Salesman A's insurance. Since every salesman would always put down his competitor then both will say that other insurance does not hold enough promise, then the wealthy client would have no choice but to deny each salesman a sale.

Looking at the market as a host of decision-makers' strategy would also lead us to look at it as a venue for conflict and cooperation, of equilibrium and disequilibrium. We have to bear in mind that the market is an impersonal entity which is home to self-interested people. Individuals who transact in the market are motivated by the wealth they will gain whether it be wage for laborers, rent for landowners, merchandise for consumers or profit for producers. In view of the limited resources which the self-interested individuals could acquire, the impersonal relationship is born out of exchanging goods and services not for benevolent reasons but in pursuit of their desire to gain and maximize what they have. Here is where the conflict lies. Everyone wants to maximize his gain but there is not enough to gain, much less to maximize. Resources are limited and everyone wants to grapple for it. Money is not abundant. Most often you either work hard for such a scarce resource or even bargain for it. But beyond this, the self-inventing individual also transacts and agrees. When a person buys something he tries to cooperate and later agrees to pay an amount in exchange for a merchandise. Even those locked in bargaining where Game Theory simulates their transactions are engaged in cooperation one way or the other. If capital is abundant then workers may be scant. But if there are too many graduates that labor overflows, then there is scarcity of capital in order to translate this human resource into economic benefit. But in all these scenarios, the amazing thing is that, through the price mechanism of the market, the limited resources are allocated. And this is what Adam Smith calls "the invisible hand of the market."

Yet even if limited resources are allotted, the market is not an arena where everyone is satisfied, where everyone brings home the best merchandise he desires, where everyone goes home happy with his abundant salary and where a capitalist closes his day with surmounting earnings. Not all investors earn profit over time for there are times when they lose part of their investment. Not all salaried people are well paid and are happy with their wages. If they are happy with their take-home-pay now, later they won't be anymore. And not all consumers are always happy with the services they receive or the commodity they purchase. Even if they buy the same brand, another brand may emerge and snatch away their preference from the old one. Information is important in market transactions but not everyone is imbued with all the information he or she needs to execute the transaction. Information that is needed to buy or sell is also scarce.

There is an irony to this. While information is vital to the execution of economic activity, we tend to covet full information of the commodity or services we desire to have. But theory would tell us that the more information we get, the more we could lose. The more complete the information we could have, the more unpredictable the market would become. This is the heart of the Efficient Market Theory that governs the market of securities with the stock market as one sterling example.

The origins of this theory can be traced from the Random Walk Theory espoused by Lois Bachelier in early 1900. Bachelier asserted that prices of stocks behave randomly and the probability of getting profit is the same probability of losing one." The idea was picked up by Eugene Fama in 1965 when he advanced the Efficient Market Hypothesis. A review and evidence of the theory was published in the *Journal of Finance* in the 1970s as the idea was also picked up by an eminent economist Dr. Paul Samuelson. The EMH, which was later on accepted as the Efficient Market Theory, asserts that efficient an financial market is one in which "security prices reflect available information and respond rapidly to new information as soon as it becomes available." (Brealey, Myers and Marcus 2004:689).

> Information is news that may affect the price which is unknowable at present but may appear randomly in the future (Fama 1995).

Information is the lifeblood of the financial market and it is information that makes the market efficient because it has the capability to make market prices to "randomly walk."

> A market where there are large numbers of rational, profit-maximizers actively competing, with each trying to predict future market values of individual securities, and where important current information is almost freely available to all participants. In an efficient market, competition among the many intelligent participants leads to a situation where, at any point in time, actual prices of individual securities already reflect the effects of information based both on newspaper-article-events that have already occurred and on events which, as of now, the market expects to take place in the future. In other words, in an efficient market at any point in time the actual price of a security will be a good estimate of its intrinsic value (Fama 1995).

A market participant or an investor in a financial market like the stock market holds millions of money in his hands which he wants to earn a lot of profit at the least amount of time. With a huge amount of money, losing becomes a very big risk. A market player would always want to be sure of his buy. But what he buys is not something that he could physically examine like apples or lemons. He will buy shares of stocks of companies with which he could only rely on reports. There are three kinds of information that he could grab: technical analysis, fundamental analysis and inside information. Technical analyses are graphs that chart the movement of the prices of stocks within a period of time. If the lines are moving up, the prices of certain stocks are rising. This is the day-to-day outcome of stock market transactions pertaining to a certain stock. Then there are fundamental analyses which are reports of the financial assets and liabilities and summary of the performance of the company selling the shares. Increasing assets and decreasing liabilities and growing investments are indicators of a good buy. Next is inside information which is a host of confidential pieces of information that could greatly affect the movement of stocks. It could be new breakthroughs which are not supposed to be circulated to the public or damaging scandals which could taint the company's profile. There is just one more addition to the kind of information and that is public information in the form of news circulated by newspapers or the mass media. It was found that the Philippine stock market index "fluctuates and vibrates depending on the mood or climate that bundles of extrinsic and intrinsic information paint" (Gabriel 2014:440).

> There are extrinsic bits of information outside of the market that affect the movement of the overall trend of the prices of most stocks. These bits of extrinsic information create a climate that paint the overall conduciveness to invest or reclusiveness to withhold investment. This research found out that the bunch or bundles of information with varied subjects or those concentrated on certain political or economic issues are significantly correlated and they possess the tendency to affect the movement of the stocks or currency market. The bunch of information creates a climate that would induce or discourage capital or money market investors in their investment... The market then is efficient because the information significant to invest is made public and no one has the capability to win outrageously excessive returns over the rest. The trading becomes a skill of bargaining and timing. Extrinsic information which this research has found to be correlated and could affect stock preferences are also public knowledge which no one has monopoly of. The climate which this bunch of information creates may induce investors to bet more of their money or deny the market of their investments. If you are picking coffee beans, the climate and season would have made the beans

ripe and the atmospheric condition of the day would have motivated you to harvest them. In fact you would have already known what to pick by simply looking at the bushes. But because it rained that day, you chose not to pick the beans (Gabriel 2012: 43).

All these pieces of information should be made available to the market and market players consider these pieces of information to be vital to execute their strategy. The strategy is to buy stocks at prices when they sink low but sell the stocks at prices when they begin to soar. Using these pieces of information create three kinds of market: the weak, semi-strong and strong markets.

The weak form assumes that prices reflect all the information in past prices. Prices appear to wander randomly, virtually equally likely to offer a high or low return on any particular day, regardless of what has occurred on previous days (Brealey, Myers and Marcus 2004: 161-162, 689). Since market players would watch and make decisions on how transactions evolve within the day, the movement of the prices of stocks provided for by technical analyses would be the reliable source of information. If most investors would do such thing, then they would be gobbling technical analysis and such information would not anymore give them extra returns. It is fundamental analysis that could give them an edge. If, investors, on the other hand, would rely on fundamental analysis coupled with technical analysis, then even fundamental analysis would not give them extra returns anymore. This is the semi-strong market. The semi-strong form assumes that prices reflect all publicly available information and share prices rapidly adjust to new public information (Brealey, Myers and Marcus 2004:689). Only confidential information in the form of inside information could make one secure some returns. But if one investor is raking huge profits at one time, other investors would suspect he is holding inside information and rather than trying to find out what he knows, other market players would simply follow his lead and buy what he is buying. Since many players are seeking to buy the stocks, he is buying then, instead of gaining extra profit, he would end up losing some profit for it would lower the price of the stocks with many investors trying to secure his buy. This is the strong market where not even inside information would give him an edge. The strong market holds that prices reflect all acquirable information (Brealey, Myers and Marcus 2004:689).

The irony behind this is that, information that most market players would like to hold on in order to secure more gains, become the thorn that would make them lose. Since all market players would have the same information, then instead of gaining extra returns, they would end up losing more. It is like placing a plate of apple pie on top of the table surrounded by hungry individuals. They know that by grabbing a piece, their hunger would be served but if all of them will do the same thing, all at the same time, then most of them will only end up with broken pieces and crumbs, not enough to satisfy their hunger. Winning becomes a purely random affair. No market player would win all the time. The outcome of market strategy becomes all the more probabilistic.

Thus even if economists try to achieve the equilibrium, the market is most of the time in disequilibrium. Yet this disequilibrium is not bad after all. And in the midst of this disequilibrium, the self-inventing being tries to fit himself in for it provides the incentive for capitalists to re-think their strategy of selling or modify their products to meet the competition with the fall out of loyal consumers who switched to other brands. Disequilibrium is good for the workers to improve themselves, work harder for promotion that seems so scarce too or change their employment in favor of another worker to come in. In the same manner the slight disequilibrium in price provides the incentive for producers to increase the supply of their produce if prices are high due to increased demand. But the influx of additional supply would lower the price to the benefit of consumers. In the wake of all these, we find the market to be dynamic. It is not characterized by a straight line, but marked with fluctuations which permit some to lose and some to win. This allows the self-inventing individual to situate himself in the dynamics of the market that allows its on-going construction and reconstruction. But the beauty of the market is the ability of these self-inventing beings to agree despite the scarce resources presented to them. The exchange of money for goods or wage for services is a way through which consensus is made and their milieu of scarce resources is constructed and continually reconstructed.

Textualizing Communicative Action

There are, however, things which economic relationship falls short of characterizing human beings. Come to think of it, if physics has tried hard to unlock the nature and structure of atomic particles, which until today has mystically boggled the curious minds of physicists, social science has also a difficult time unlocking the nature of human beings. While postulating humans as self-interested maximizers of resources, such rationality is not enough to fully describe their multi-faceted nature. Consider this example.

Let's say an apple pie was given to two individuals over dinner. Person A had not eaten his lunch and so was hungry. Person B had a full breakfast and lunch. For economics it would just be rational that Person A would strategize to get more than half of the apple pie, assuming that he is a rational calculator of the resource presented to him. But suppose, instead of having a bigger share, Person A and B talked it out and split the apple pie in half. Person B is now at an advantage over A and would be presumed to be rational having acquired more than what he needs than Person A. But is Person A irrational after all, having given up more portions though he was hungrier than the other person?

This question leads us to the problematique which Habermas sought to address. Habermas belongs to the critical school of sociology which has its origins at the Institute of Social Science Research at Frankfurt University in Germany. It was staffed with the leading German intellectuals of his time like Theodore Adorno, Max Horkheimer, Herbert Marcuse and Friedrich Pollock. They maintained their association even at the time of growing Fascism in Germany and Stalinism in the former Soviet Union. They were drawn together by a common interest toward their commitment to a "moral concept" of progress and emancipation framed with the idea that knowledge should be put to use in order to achieve a just and democratic social order (Farganis 1993: 335).

Habermas studied under Adorno and imbibed the critical school's doctrine that a theory should not only explain but also change. But he later dissociated himself from the institute when he engaged them in an intellectual tussle over his criticism that they were undermining the communicative aspect of human relationship. The major focus of Habemas' work is the survival of democracy in a society which is marching on towards scientific and technological development. He argued that Max Weber's "Iron Cage" description of modernity is unnecessary if we could emancipate individuals from the colonization of a functionally rational system which results in the loss of individual's liberty. Weber realized that an irony exists in a democratic society which subscribes to the idea of espousing liberty and creating relationships only to constrain an individual. A factory, for example, is a rational system created towards the maximized production of goods at lower cost but the system seems to constrain individuals who would be forced to do repetitive work over a period of 8 hours a day. Thus it seems that despite a rational system built, an individual is condemned to irrationally lose his personal choice to do what he wishes within the 8 hours.

In order to address this problem, Jürgen Habermas postulated his Theory of Communicative Action. He argued that there is not just one kind of rationality which the economic calculator of resources would always be presumed to have. Remember Person A and B who had to split the apple pie. For the economic individual, Person A who had not taken his lunch and would be presumed to be hungry would be said to have behaved irrationally by agreeing to acquire the same amount of apple pie as person B who had just taken his lunch.

For Habermas, the Theory of Communicative Action recognizes that there are two types of action that necessitates two kinds of rationality. The first is instrumental or strategic action which can be understood as following technical rules and can be evaluated in terms of its efficiency in dealing with the physical world (Roderick 1986: 109). This type of action is oriented towards successfully dealing with what the physical world offers. This is the realm of the economic person's rationality. The efficiency of his act depends on the strategy on how he could maximize the use of his resources to attain his objective. But if two people agree, even if one would find himself slightly economically disadvantaged, would he still be considered rational? In this instance, Habermas advanced the concept of communicative action.

> Communicative action occurs when social intercourse is coordinated not through the egocentric calculations of the success of the actor as an individual, but through the mutual and cooperative achievement of understanding among participants (Habermas 1984: 285, Roderick 1986: 101).

The central theme of this theory is action oriented towards reaching an understanding in the direction of achieving communicative rationality.

> The concept of communicative rationality is connected with the ancient conceptions of logos and carries with it connotations based ultimately on the central experience of the unconstrained, unifying, consensus-bringing force of argumentative speech, in which different participants overcome their merely subjective views and, owing to the mutuality of rationally motivated conviction, assure themselves of both the unity of their objective world and intersubjectivity of their lifeworld (Roderick 1986: 113, Habermas 1984: 10).

Thus we find that individuals, who have reached an agreement over a certain matter, though for an instrumental rationalist they may have been disadvantaged, have in fact, attained communicative rationality. But how is communicative rationality reached? Here communicative competence is needed. Habermas claims "speech acts are themselves social actions" (Roderick 1986: 111). Language is analyzed as to its social use. Likewise, communicative action, with its goal of reaching consensus, would need to achieve four elements which Habermas calls validity claims. Habermas assumes that competent speakers as social actors in a community have the ability to distinguish these elemental validity claims within the spheres from which they emerge (Roderick 1986: 110).

One of these elements is truth. This is a validity claim directed towards external nature. The speech act is directed to an objective evaluation of the world disaffected from the social actor. For example, if we go to the seashore and taste the sea water, all of us will claim it is salty. And it will stay salty even if others will taste it. Claiming otherwise is untrue. If truth is a validity claim directed towards external nature, rightness is a validity claim directed towards society where speakers evaluate the rightness of certain speech acts in accordance with the rules governing social relations. Let's say we are already on the seashore and discussing whether to dump garbage to the sea. Even if we concede to do it, such an act does not conform to the rules of good and moral behavior towards others and towards the environment. Beyond these claims, a speaker also needs to be sincere. This is another validity claim directed at the speaker's internal nature. It is an expression of one's inner desire, wishes and needs which raise a claim in regard to his sincere disclosure. What if instead of dumping garbage into the sea two individuals agree to clean the seashore within their reach. But the question then would be, are they sincere on their resolve? Lastly the speech act should consist of the final validity claim directed at language itself. This means that the speech act should be comprehensible where the language used should conform to grammatical correctness so it could be understood (Roderick 1986: 110-111). Communicative competence then would mean that a communicative act should contain claims valid enough to contain truth, rightness, sincerity and comprehensibility in order to reach a consensus and achieve communicative rationality (Figure 5.13).

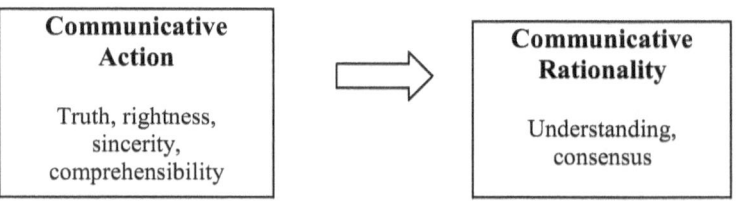

Figure 5.13 Theory of Communicative Action in a Diagram

Habermas considers the attainment of consensus as an outcome of communicative action. Consensus then is communicative rationality in action. The theory then would preclude that communication oriented towards understanding can be attained only to the degree by which participants credibly sustain these elements into claims a) that the propositional content of the speech act is true, b) that it is socially appropriate, c) that it is sincerely conveyed d) and that the utterance is comprehensible. It would be noticed that Habermas does not exclude the objective world as a source of discourse or speech act. In fact, he includes the truth-value of theoretical discourse that gravitates on objective truth, aside from the normative appropriateness and sincerity of utterances.

All along, the Theory of Communicative Action has centered on the speech act but there is a concept which it still needs to address and that is the speaker whose background could constitute the nature of the act. Here Habermas introduces the idea of the lifeworld.

> Habermas defines the lifeworld as the context for language which stands behind the back of each participant in communication and supports the process of understanding. Every actual consensus is achieved against this uniquely pre-reflective form of background relations... The lifeworld is a resource for what goes into explicit communication which can become subject to criticism. The lifeworld itself, however, always remains implicit, pre-reflective and pre-critical (Roderick 1986: 119-120, Habermas 1984: 70-71).

The concept of the lifeworld can better be understood by contrasting it with the social system. Social systems are organized structures which follow functional rules and where instrumental action takes place. The lifeworld, on the other hand, provides meaning and definitions that are drawn for social reproduction (Roderick 1986: 120).

> For Habermas, the lifeworld depends upon the social system, both in terms of material production

(the economy) and organization (state). The social system depends upon the lifeworld for both the reproduction of socialized individuals and the continuation of coherent cultural traditions (Roderick 1986: 131).

The lifeworld stands as the background of every communicative act. Habermas argued that the outline of one's communicative act is shaped and reproduced in his lifeworld which becomes the fusion of his objective world (truth), subjective world (sincerity) and intersubjective world (rightness).

If we go back to our illustration before, the Theory of Communicative Action does not make Person A, who has given up a substantial portion of the apple pie to Person B though he is hungrier; irrational than the other guy as long as they have reached an agreement to divide the pie equally. In the light of this theory, the self-inventing nature of human beings becomes well situated in a milieu they could well construct through communicative action. The consensus they build is evidence of communicative rationality. Agreement builds and rebuilds their milieu through communication and in the same manner creates and recreates themselves.

Postscript 2

The milieu-under-construction then is a pattern of constantly emerging realities. This is a milieu that is continually being built or reconstructed. Theories do not only textualize the milieu by presenting its patterns but also rebuild it in the process. While theories capture these patterns, disassembles them into conceptual parts and assembles them back to an understandable whole, the theory in turn affects the patterns by reconstructing them. Take Marx's Dialectic-Historical-Materialism. The theory explains the relationship between the labor class and the bourgeoisie in view of the political and legal infrastructure built by the owners of the means of production. He argues that the contradictory relationship of the two classes becomes fetters to the labor class, which will have no option but to change the political and legal infrastructure by claiming the ownership of material production. This theory does not only present and explain the patterns of social and historical progression of society but the end result that the theory hopes to accomplish is to overhaul the pattern of the social system.

The same is true with the market which economics textualizes. Human agents are not only rational and self-interpreting but also self-inventing. The self-inventing character of human beings fits in to the interactive pattern of the market marked by equilibrium and disequilibrium. The buying and selling as characterized by the preferences, strategies and the availability of limited resources create a continually emerging, reconstructed system. The market works within the dynamics of the quantity of demand and availability of supply as intervened by the price. But without the bargaining, negotiations and eventually agreement among buyers and suppliers the market will not work. The negotiations and eventual consensus make for the constructive character of the market. Habermas' Theory of Communicative Action evidently explains the interactive mode of human beings engaged in speech acts which in turn recreates the resulting structure of their communication by arriving at a consensus.

On the other hand, theories can also capture and textualize the reconstructive process of certain phenomena. The mitotic and meiotic processes coupled with the crossover and recombination of chromosomes are processes that produce and reproduce life which the Chromosomal and Molecular Theory of genetics could account for. In the same manner, the Quantum Theory and Heisenberg's Uncertainty Principle explain the probabilistic nature of a particle's location and momentum in the minute scale that determining where it is and how fast it is going at the same time is impossible given the fact that the observer could disturb the particle at the moment of observation. While it is assumed that this is the property of matter in the atomic world, it brings to fore the reconstruction of the particle's history with the observer himself being part of it.

All these describe the milieu-under-construction. Human beings construct and reconstruct their social world and the constructive and reconstructive patterns are textualized just like the reproduction of cells and the mechanics of atomic particles. It is not just the nature of these theories to be reconstructive but the milieu-under-construction itself is reconstructive, where patterns emerge, realties are experienced and the milieu is created and recreated. Some aspects of this milieu is not only reconstructive but also probabilistic. Statistics governs the nature of an electron's movement, probability describes the transmission of hereditary traits. But human interaction cannot simply be deduced into equation. While the biological process of cell and the physical mechanics of nature can be captured by closure, participation in social action in its naturally occurring environment can be captured in human interaction. There is a downside to this, however. Since human beings are self-inventing, the observer can affect the social interaction and, in turn, bring about the recreated human interaction, making him part of the observed phenomenon. Yet we find this to be a direct evidence of the milieu of emerging realities, a milieu that is continually being constructed and reconstructed... the milieu-under-construction.

What theory does to a milieu-already-constructed is the same as what it can do to a milieu-under construction. Theories are able to disassemble the phenomenon into parts, screw them back into their relationships and provide a knowledge of what the occurrence would be once these elements function as a whole (Table 5.1).

Table 5.1 Summary of Theories in their Deconstructive, Reconstructive and Foreconstructive Mode in Milieu-under-construction

Theories and Phenomena	Deconstruction of the phenomena into parts	Reconstruction of the parts into their relationship	Foreconstruction of what occurs
Quantum mechanics Radiation and movement of particles at the sub-atomic level	Energy (E), Wavelength (λ), Planck's constant (h), Velocity (V), Location (L), Wave function (λ)	Radiated energy (E) of a wave is directly proportional to its wavelength (λ) and Planck's constant (h). At the sub-atomic level, the location (L) and velocity (V) of particles cannot be known at the same time. Sub-atomic particles in movement act both like particles and waves creating their wave function (λ).	Radiated energy exhibiting waves like light is propagated in granules or quanta. Electrons like granules that circle the nucleus behave like that of a cloud, if its location is detected, its velocity cannot be known but if one knows its velocity, no one would know where it is.
Theory of Heredity Transmission of character traits to the offspring	Dominant trait (A), Recessive trait (a)	The possibility that dominant trait would show in the offspring is ¾ or 75% while the possibility that recessive trait would show is ¼ or 25%. Genes transmit these traits.	Parental traits that greatly manifest in their offspring are the dominant trait while those that rarely show are recessive.

Theories and Phenomena	Deconstruction of the phenomena into parts	Reconstruction of the parts into their relationship	Foreconstruction of what occurs
Strand Separation Theory DNA replication	DNA, DNA polymerase	The purine-pyrimidine strand of the DNA begins separating into Y strand with the addition of a complementary nucleotide through an enzyme, DNA polymerase.	The cell divides and multiplies. Exact copies of the DNA are produced with the addition of an enzyme present in the cell.
Game theory Strategic decision-making	Pay-off (P), Strategy (S)	Rational individuals would prefer a strategy (S) that would gain them the highest pay-off (P) or minimal loss in economic transaction. But depending upon the game, they may win or lose everything in a Zero Sum Game, would win some or lose some in a Non-zero Sum Game, or would minimize what they could lose in a Prisoner's Dilemma Game.	When confronted with an option a rational individual would always prefer to gain the most but since other rational individuals are also thinking of the same thing, they would end up doing the same thing, either getting everything or splitting the gain between the two or minimizing the loss that they could incur.
Efficient Market Theory Financial market in operation	Information (I), Strategy (S)	In order to acquire returns in the financial market, players would opt to have all available information (I)	Instead of gaining more than the other market players, investors armed with the same information would end up not

The Milieu-under-construction

Theories and Phenomena	Deconstruction of the phenomena into parts	Reconstruction of the parts into their relationship	Foreconstruction of what occurs
		made public and execute their strategy (S) of "buying low and selling high."	gaining as much, making the market operate randomly.
Dialectic-Historical-Materialism Political economy and emancipation of the working class	Social Class (C), Opposition (D), Productive means (P)	The development of economic systems from feudalism, capitalism to socialism is marked with a social class (C) owning the productive means (P) which produces an opposition (P) from another social class. The conflict ascends another economic condition or system.	If workers would improve their lot, they would have to own the productive means which is owned by capitalists who have interests in opposition to them.
Theory of Communicative Action Rationality and consensus building	Communicative rationality (Cr), Agreement (A), Truth (T), Rightness (R), Sincerity (S), Comprehensibility (C), Communicative Action (Ca)	Rationality does not always mean maximizing gain and minimizing loss. The attainment of agreement makes individuals rational or communicatively rational (Cr), if the communicative action (Ca) or speech act possess truth (T),	Individuals may not always achieve or maximize their gain but if such gain or loss is a product of an agreement, then they are rational or communicatively rational. Agreement is reached if their transactions or communicative

Theories and Phenomena	Deconstruction of the phenomena into parts	Reconstruction of the parts into their relationship	Foreconstruction of what occurs
		rightness (R), sincerity (S) and comprehensibility (C).	actions possess truth, are comprehensible or understandable by all parties, are morally right and they are sincere in their intentions.

If theories can textualize the milieu-already-constructed, they do as well in the milieu-under-construction.

Chapter 6

The Milieu-of-the-text

Nature, biological life and human relations operate in both milieu-already-constructed and milieu-under-construction. The physical and social worlds work in the two milieus. Mechanics in nature obeys deterministic laws at the level of the immense but behaves in probabilistic, observation-interactive manner at the level of the sub-atomic. Biological life is uniquely built with specific molecular components and governed by discoverable principles in the transmission of hereditary traits but its reproduction, where hereditary traits are replicated and transmitted, is characterized by probability and randomness. Human relations in the area of economics also operate on principles that are quantified by the cost and amount of transactions but it also works through the construction of humans adopting themselves to certain theoretical precepts.

What theories do is capture the operations of some specific aspects of nature, biological life and human relations and textualize them. Theories are texts. They are intersections of the milieu that is already constructed and the milieu that is under construction. The world is also an aggregate of realities and these realities are interpreted experiences which make the world an experience. Theories, therefore, reconstruct the world already built. The world, therefore, is mastered through our textualization of it and theories that are the products of our textualization construct and reconstruct our perspectives about the world. Inasmuch as human beings device theories, theories also make them as well. And since theories are the heart of science, science is nothing but the textualization of the world. And since theories are human inventions, science also is a human invention. There are just a few items which this project has to iron out.

Since the world is an aggregate of realities and realities are interpreted experiences, then anything which is not yet experienced is not real and thus not part of the world. To clear this out, we have to answer this question: Is there really a world out there waiting to be discovered? Is there really a world that exists outside of the observer? When Galileo peered through his telescope, did the lenses manufacture the light that Galileo observed or was the light already there that the lenses of the telescope just captured the rays emitted by the planets he was observing? The same question would be posed to Edwin Hubble when he found that the galaxies are moving away from each other. Even without Galileo or Hubble, even without the scientists to discover them, even without telescopes to aid their discoveries, these planets would be in their exact position and behaving the way they do. Likewise, even if there were no microscopes to investigate the cell or no scientists to probe into the essential unit of life, the cell and the processes of its division and replication would still proceed. The only thing is, had there been no telescope or microscope, had there been no instrument to investigate the objects and had there been no scientists to bother themselves with these, these objects would not have impinged on our senses and would not have been axiomated. And so before that time, they are not real to humans but they exist there waiting to be realized. And if they are experienced, they still await interpretation, so they could be fully comprehended.

For the milieu-already-constructed, the world already exists. It only has to be made real through our senses and through our interpretation of them using axioms. But for the milieu-under-construction, it would be quite a different story. Since this is a world of emerging realities, the world that is not yet experienced has not been or has not become. It is not yet constructed, therefore, there is nothing to experience. So this world does not exist. It would be interesting, however, that even with the pre-existing world that is already constructed, humans would still interpret and reconstruct what they discovered. Here the milieu-under-construction takes place. Thus with the discovery of some aspects of the existing world, its interpretation and translation into rules ensure an emerging reality which would then be interacted upon. Yet there are certain points which need clarification. In the reproduction of biological life through mitotic and meiotic processes, the mechanism of life exists even before observation takes place. But what makes for the constructive nature of biological life is not in the interaction of the observer and the observed like quantum mechanics but in the mechanism through which life is reproduced. The cell and its process of division are already predetermined that exist even before discovery and observation. But the mechanism of cell division is in itself constructive and reconstructive. How the cell builds and reproduces itself is already constructed that await discovery. But once the process goes, by itself, the reproduction and rebuilding become an emerging reality. For quantum mechanics, however, there is another unique story.

Quantum Disciples

What makes quantum mechanics part of the milieu-under-construction is the emergence of a reconstructed condition as the observer interacts with the particle under observation through the electromagnetic waves beamed on it. In fact, Heisenberg's Uncertainty Principle and Schrodinger's Wave Mechanics are universal principles fit for a milieu-already-constructed. Nevertheless, the effect which observation and measurement does to the particle being observed brings about an emergent reality which the milieu-under-construction could characterize.

It has been previously discussed that the subatomic particle behaves in probabilistic manner, which would make it impossible to ascertain the velocity and location of the particle at the same time. Thus the electron that moves around the nucleus cannot be ascertained where it is, if its velocity is known, and if it has been located, how fast it is moving cannot be determined. Since Planck has postulated the granular nature of energy depending on the wavelength of emission, confirming Planck's Quantum Theory was Einstein's Theory of Photoelectric Effect that verified the particle-like behavior of electromagnetic waves. Waves then behave like particles and particles act like waves. The peculiarity of quantum mechanics led Schrodinger to formulate the Wave Mechanics of what these particles obey which also brought him to devise an analogy that has adopted his name. This is the Schrodinger's Cat.

Schrodinger's analogy is actually built on Heisenberg's Uncertainty Principle which presupposes that observation has caused a disturbance in the behavior of the particle, that simultaneously locating it and determining its velocity at the same time are an impossibility. The more precise a researcher who wants to determine either of the variable would make the observer use higher frequency of waves at higher energy, thus disturbing the particle even more. The wave that was intended as an instrument of observation has become a component of the object being observed. With

The Milieu-of-the-text

the wave equation that Schrodinger formulated, his analogy goes like this:

> "A cat is in a box and inside it is an apparatus where an electron passes to switch on a device that breaks a glass of poison. There are two possibilities then. If the glass breaks, the cat would be dead, if the glass remains intact, the cat survives. That would mean, if the electron behaves one way to trigger the apparatus, the cat dies for the glass would be broken. But if the electron behaves the other way, the cat remains alive. Since the electron behaves in several possibilities, then we would not know if the cat inside the box is dead or alive. Quantum mechanics which accounts for all possibilities would say that the cat is both dead and alive. Only upon opening the box after the electron passes through the apparatus would one be certain if the cat is really dead or alive. By that time, observation has "collapsed" to only one possibility and determining what really happened if the electron behaves one way or the other carries with it the inclusion of observation" (Kramer 2013).

The influence of quantum mechanics was so great that other physicists joined the bandwagon but there were those who rejected its consequential predictions. The problems with quantum mechanics were the issues of measurement and observation, as well as, the nature of the objects being observed. The fact is that the atom cannot be seen even through a powerful microscope because it is much smaller than a quantum of light. If the atom cannot be seen, how much more its subatomic particles? The only way to observe it is to disturb it. Thus without the disturbance, the particle cannot be observed. In this regard, quantum theory has generated 3 kinds of individuals.

There are those who are pessimists. Like Einstein, these scientists believe that quantum mechanics is incomplete (Kaku and Thompson 1987: 48). Einstein believes that deterministic laws that underlie the movement of subatomic particles still await discovery and the probabilistic description of its behavior illustrates only the interaction of the particle when observed. He believes that this is not the nature of subatomic particles even when not observed.

The optimists, however, aim their big guns on the pessimists. One issue in point is the result of the double-slit experiment. When light is beamed on a double-slit partition, the reflected light forms fringes of bright and faint light. This proves that light behaves like waves, where at certain intervals, the light waves merge to create bright reflections, while at other fringes, the light waves cancel out to create faint and dark reflections. This confirms the wave character of light and, coupled with its granular nature, light is then both wave and particle. The remarkable thing, however, was when this experiment was repeated using an electron beam, the same bright and dark fringes appeared. It would be presumed then, that one electron passes both slits at the same time (Hawking 1988: 63). This would illustrate the wave-like property of subatomic particles. Thus as electrons move with definite speed around the nucleus, its motion would not be described in definite lines like planetary orbits but its movement would appear like a cloud around the nucleus. In this manner, the statistical description of its location and motion would corroborate the wave function of the particle. The optimists, therefore, believe that the probabilistic behavior of the subatomic particle is its fundamental nature. This would mean that at the realm of atoms, determinism breaks down and nature behaves probabilistically.

The Milieu-of-the-text

Then there are the pragmatists who subscribe to the assumptions of quantum physics and extend the theory to other areas or discipline. Stephen Hawking could be classified under this category when he postulated the "No Boundary Proposal." Since Einstein assumed that everything in the universe moves, then it is, in fact, expanding. Edwin Hubble in 1920 observed that galaxies are moving away from each other, confirming the universe's expansion. If we would therefore want to know the condition of how the universe began, then we could just logically reverse the expansion, until everything is compressed into a singular particle. This is the singularity condition of the Big Bang Theory. But Hawking objects to this in effect. If the universe started out as a particle then quantum mechanics operate. Einstein, who germinated the idea of an expanding universe as his Theory of General Relativity would suggest, cannot just accept that at the atomic level, indeterminacy would result. If the universe is compressed in reverse, a particle would result. Einstein's theory is deterministic but at the atomic level, the particle behaves probabilistically as quantum mechanics posits. Einstein then objects, "God does not play dice." Hawking's objection results in his solution. Since the fundamental behavior of the particle is probabilistic as Quantum Theory posits, then at the singularity level, there would be no point particle but "a smear, a cloud or fluctuation."

> The boundary condition of the universe is that it has no boundary. The universe would be completely self-contained and not affected by anything outside itself... Without singularity, then the universe would have no boundary. Without boundary it would have no beginning. It would neither be created nor destroyed. (Hawking 1998: 144).

The tide that quantum mechanics generated had also created ripples in the philosophy of science that generated the extension of its principles to other aspects of nature. Along this line, a philosophical puzzle is posed, "when a tree falls in the forest, does it make any sound if there is no one to hear it?" (Kaku and Thompson 1987: 46) Since there is a need for a consciousness to observe the phenomenon, what if there is no consciousness to witness it? Had the phenomenon taken place? There would be different answers to this question, depending on the belief one espouses. For a pessimist who views this puzzle within the frame of a milieu-already-constructed, then the sound of a falling tree would have surely reverberated even if no one was there to hear it. For an optimist, who views this within the milieu-under-construction, then there was no sound, since there was no consciousness to construct it on his own. But for the pragmatist who is himself a quantum disciple, then it is both, there was no sound and there was sound. It would just need another event to make the scenario collapse into one condition.

Science is Disturbance

Since quantum theory has unearthed the controversy regarding the issue of observation and the real world, the debate between different schools of the philosophy of science becomes paramount. Realism assumes that theories can be judged on the truthfulness of their claim depending on their correspondence to the natural world they seek to explain. Realism caters to the correspondence theory of truth, which argues that the world does not depend on what is said about it. This would mean that the world has properties which stand on one hand while theories also have their own characteristics which stand on the other. The amount of truth which theories contain would depend on their correspondence with the properties of the world. This is challenged by anti-realism of the empiricist camp. Anti-realists claim "science is not trying to describe the way the world really is but is only attempting to account for the available empirical evidence" (Rouse 1987: 128). This would mean that theories can only be proven for their truth-value only up to that point where empirical evidence is available to verify them. On this note, things that cannot be observed would hardly be included within the province of science since they cannot be observed and correspondence theory applies only up to that point where empirical evidence could match the claim. On the farther side of realism is constructivism. It assumes that everything in the world is "theory-laden" (Rouse 1987: 128). The world cannot be understood without the vocabulary our scientific theories ascribe to them. Constructivists outrightly reject correspondence theory. As such constructivism argues that theories are not tested based on their correspondence with the world but in competition or comparison with the claims other theories uphold.

These three schools of thought, however, agree that objectivity is the underlying theme beneath all scientific efforts. The observer is disaffected from the object of study. The object is investigated from a "distance." The distance is not actually geographic, but it refers to the entire procedure which should be free from the interaction or contamination of the observer. Such interaction may lead to flawed results for it is assumed that the object of study exists in a world apart from the investigator. Thus all the three schools of thought claim to be searching for truth or objective truth. The truth would necessitate its translation into theory-statements. But the theory has a system of its own. As made up of statements, it is language with distinct vocabulary that would hope to describe the object with precision. Precision becomes a dividing issue among the three schools of thought. For the realists, precision is determined by direct correspondence. It would presuppose that the truth content of a theory is measured by how much it corresponds to the object being studied. The empiricists are more modest, claiming that only those observables could be theoretically constructed. This would be tantamount to saying that those which cannot be observed, cannot be verified. It is a region where theory has no territory. The constructivists, on the other hand, believe that the theory's precision upon which verification rests can only be gauged, not through correspondence with the nature of the object, but through its examination with other theories and how it could stand up with its claims once confronted with them.

This present project has certain objections to these claims. First, theories do not seek for truth or objective truth. If these philosophies of science or even scientists proselyte that objective truth is the end of their scientific pursuits, then objective truth would, in fact, be a contradiction. Objectivity means detachment of the observer from the object of observation. This I would call absolute objectivity, which in science, is impossible to attain for an absolute detachment from the object being observed will result to not doing anything at all. Absolute objectivity is complete detachment which would result to zero activity leading to zero disturbance and in the end resulting in zero science. What science can achieve is not absolute objectivity but instrumental objectivity. This type of objectivity uses an instrument to detect and measure the object being observed. It could be a form a measuring stick, a thermometer, a Geiger counter, a survey form or an interview guide that permits the instrument to identify and calculate, removing the scientist from any subjective inclination. It makes identification and calibration standard. In a scientific inquiry, the procedure of investigation and instrumentation are critical markers where objectivity rests. But the scientific enterprise is a pursuit of intervening in the world. It is an endeavor characterized by disturbance into the object of examination. Experimentation is disturbance. If you put a mouse in a cage and observe its behavior, then

The Milieu-of-the-text

whether you are a student of science or a full-blown scientist, what you did was to disturb the mouse's world. Worse is if you will inject a foreign substance into the mouse and observe its effects. The next recourse would be to observe the object in its natural habitat. But even by observing gorillas or hippopotamuses in the wild, the presence of the scientists would somehow disturb the animals in their habitat. In the same manner, to observe a cell under a microscope necessitates the killing of the cell and dyeing it; thus disturbing it. Shooting an electron through a double-slit partition is disturbing its natural motion and observing its behavior. Even Galileo's experiment of rolling two balls of different masses on a slope is disturbing the two objects in order to ascertain its movement. Determining the chemical composition of certain compounds through electrolysis also disturbs the way elements are bonded together. For the social sciences, the questionnaire as an instrument which aids in achieving objectivity in a survey is also a device that elicits disturbance in the research area. A questionnaire forces an individual to stop what he or she is doing and pour in his undivided attention to the questions, thereby disturbing his or her world. The respondent might even provide answers which he presumes the researcher would wish to gather in order to prove his hypothesis. Even the presence of a researcher on the site, who observes, asks questions and participates in the social activity, is disturbance enough in the object of study.

There would only be a difference in the way astronomers peer to heavenly bodies through their telescopes. The bodies in space being so distant are not within the range by which they could be disturbed during observation. But what is disturbed? It should be noted that when scientists observe these bodies in space through the telescope, what their huge lenses capture are the traces of light these bodies emit. That being the case, the disturbance comes through the distortion of the light they reflect or emit. The distortion of the light that passes through the lens brings about the magnification of the object in the process. These rays of light are then captured and their images appear. If the traces of light are passed through a prism, then disturbance is accounted for by breaking down the emission into different spectra.

Doing science then, is a matter of disturbing the world or recreating it through intervention in order to observe how it operates. But if scientific pursuit is a matter of intervention and disturbance, then precision comes in by how minimal the disturbance occurs so it would not greatly affect the result of the scientific effort. For large bodies, a little amount of disturbance would not significantly affect the entire result, unless the system is "chaotic" that fractional errors in calculations or very minute deviation in a series of iterations would immensely affect the system and produce unpredictable outcomes. But for the very small, the disturbances become part of the system that measurement and observation become an integral component of its behavior.

Disturbing the system in order to investigate it, however, is done with the intention to manipulate it. Manipulating the system comes with it the hope of replicating the occurrence. But the replication of the occurrence carries with it the replication of its manipulation or the replication of the disturbance. Experimentation is a method where the object of observation is disturbed, its world recreated with the intention of investigating its behavior and replicating the entire procedure to test for consistent proof. But in the process of investigation and observation, the search for patterns and regularities carries with it the observation of the disturbance as well. If another scientist would like to confirm the previous experiment, he would recreate the entire experimental procedure, recreating the disturbance and repeating the manipulation, in order to find out if the same results are achieved. Thus the replication of manipulation carries with it the replication of the disturbance no matter how small. Now if the desire of science is the search for objective truth, then even with the strict, disciplined procedure, scientists would still disturb the system they are examining. Though how disaffected the observer is from the object of observation, as hoped by the scientific method, the observer would still have a hand in disturbing the system being observed. Thus with the results achieved comes his own disturbance. How objective would that be then? How objectively truthful would that be? Or how "truthful" would be the claim? Or rather should we ask: how instrumentally objective would that be? Or how instrumentally truthful would be the claim? Science and scientific pursuits then do not search for truth but what they seek is reality. And there are three kinds of reality that this book posits. Science always ends up with the "least amount of disturbance" (Ridley 2001:18).

But is it only a matter of the "least amount of disturbance" that science works? Or is it also knowing "how much was the disturbance?" If we know how much we disturbed then we could know how much we could eliminate, in order to find out it pristine order. Doing science would not only mean least disturbing for there are scientific pursuits that necessitate large amounts of disturbance. Take the sub-atomic particles. The way for physicists to understand the composition of sub-atomic particles is to break them by smashing them with another particle. The greater the energy of the smash, the more that the particle breaks, the more components of the particles are known. This would mean more disturbance and the greater the disturbance the greater the possibility of success to know the constitution of the sub-atomic particle. But there is also a paradox even to just knowing "how much was the disturbance."

The Challenge of Quantum Philosophy on the Practice of Science

Quantum theory then suggests that observation and measurement assume a certain degree of disturbance which becomes part of the observed. If uncertainty is the fundamental essence of nature at the quantum level, then possibilities will emerge once disturbance is committed in order to observe it. The electron in order to be observed has to be disturbed. Light in order to be propagated needs electrons to be vibrated and disturbed. But observation assumes the nature of the observed. Disturbance blends with the nature of the observed and reconstructs it. Disturbance and reconstruction become part of the observed. If observation and measurement are elements of the practice of science, then disturbance and reconstruction are part of the scientific enterprise as well. Instead of the scientist being a detached observer, he himself has become part of the observed because what he does is to disturb and observe, reconstructing the observed in the process.

Scientific discovery in the natural sciences has been positivistic in orientation. Since nature "lies out there" detached from the observer and since nature has universal essence, then the discovery of laws is necessary to give account of how nature works. These laws, however, are independent of the ones who discovered them. But quantum theory also possesses the reconstructive essence of nature, as disturbance is brought about by observation; measurement becomes part of the observed. This nature could question the practice of science itself.

A scientist is an agent. He or she is supposed to be an agent detached from the object of observation. But though detached, he is in the field of observation with the observed. Thus as an agent, a scientist is an outsider in terms of the observed but he or she is within the field of the observed.

The Milieu-of-the-text

There are, therefore, three elements here: the *agent,* the *observed,* and the *field.* The observed in the field has its own innate dynamics. A scientist, as an agent, is an invader in the field of the observed. The agent is armed with a methodology which is his tool of invasion. His or her objective is to unlock the dynamics on how the observed works. Being an invader, the scientist disturbs the observed in the field through his or her methodology.

The invasion comes in varied types: *natural, pseudo-recreative, recreative,* and *manipulative. Natural invasion* is done when the agent invades and observes the object of observation in its field without changing the field or even manipulating the subject of observation. The agent can also recreate the field and transport the observed in the recreated environment in order to observe it. This is the *recreative invasion.* On the other hand the observer can manipulate the subject itself in order to observe it. This is the *manipulative invasion.* In the *natural invasion* the agent's methodology does not change the observed nor the field. In the *recreative invasion*, the agent changes the field but not the observed. But in the *manipulative invasion*, the agent himself, changes the observed.

These types of invasion are evident in the natural or biological sciences. But in the social sciences, the agent can also do the natural invasion by simply invading the social group of the human beings as they live and act in the natural and social habitat. This is the *natural invasion.* The social scientist can also make a questionnaire and invade the subject in his natural setting and elicit answers from such instrument without changing his field or invading the subject. The social scientist still tries to preserve the field, be it natural or social, only that his social or intellectual focus is disturbed to extract answers from the subject. This is a kind of *pseudo-recreative invasion.* An experiment using human test subjects by placing them in a new environment is engaged in a *recreative* invasion. While physically or psychologically invading the human subject by injecting a substance into him, taking out something from him or by making him undergo a new intellectual, psychological, physical or social experience, the scientist is involved in a *manipulative invasion* of the human subject.

A scientist whose methodology calls for natural invasion is one who does field observation. An example would be a primatologist who is observing how primates behave in the wild. He goes into the forest and observes the primates in their natural habitat. An astronomer who peers into the sky through his equipment is one who also goes to the field and through his equipment communes with the stars in the sky to observe them through his equipment. Though an astronomer does not disturb the stars but what he disturbs are the waves of light which those stars emit as they pass through the lens of his telescope or in the prism of his spectrograph. A scientist, however, who engages in a recreative invasion is one who takes his subject from its natural habitat and observes it in a new environment. A mycologist who takes a sample of fungus and cultures it in a Petri dish in the laboratory has changed the field in order to observe the fungus, therefore initiating a recreative invasion. But a toxicologist who wants to ascertain the potential cancerous effect of certain substances on an organism would inject a test organism with the toxin and observe its effect. A proton may also be taken from its location from the nucleus, sped up in a certain direction in the cyclotron and smash the proton of another particle, thereby invading the subject itself. With this a scientist has undertaken manipulative invasion of the observed. But in all these activities, invasion is a disturbance to the object of observation.

It would be evident that scientific discovery would not succeed unless the invasion of the agent is initiated. The success of scientific discovery, therefore, depends on the

The Milieu-of-the-text

success of invasion. In the process of scientific discovery and progress, consistency of results is a landmark of the scientific enterprise. It can only be judged that a scientific endeavor has gained success if experiment after experiment the same results are obtained either by the same or a different agent. But the same findings are achieved because the same invasion has been initiated. Science acquires its success through the consistency and replicativeness of its findings. But the process of invasion is also replicated thus the same results are obtained. Invasion therefore is inherent in the methodology of the agent. There can never be a method without the agent's invasion. But since science cannot do away with invasion, what quantum philosophy challenges science with, is the reconstructive nature of the scientific enterprise.

Quantum philosophy states that observation and measurement assume the nature of the particle-wave that its velocity and location cannot be known at the same time. For quantum physicists, uncertainty characterizes the nature of the atomic and subatomic particles. The amount of uncertainty, however, presents a degree of reconstructiveness. If science cannot do away with invasion, science cannot also do away with reconstruction. What comes in along the way is the nature of the observed as to its reconstructiveness.

The observed as the object of scientific discovery can have a certain degree of inventiveness. Human beings are highly self-inventive by nature. This is the reason why the social sciences have some difficulty unlocking universal laws that govern human behavior. Human beings have a way of inventing themselves at will in the face of observation. Human subjects while answering a questionnaire as a survey instrument could alter their responses depending on their goal whether to please the observer, rig their answers to make the hypothesis work, or simply answer depending on their individual whim. Thus, instead of discovering the richness of universal laws that govern human behavior and social life, a rich variety of theories and methodologies have been formulated by human beings themselves to understand human behavior and social life. And there would be many more in the offing since the human subject is inherently self-inventive. Thus, though a social scientist makes sure about the reliability and verifiability of his instrument to account for human's social life, the reconstructive nature of the whole practice would still be existent in the process directed to the human being as the subject of observation with the scientist as an agent and his process of invasion ingrained in the methodology of the practice. Even with the desire to eliminate the threat to internal and external validity, the scientist is still an agent who invades the social world of the observed and reconstructs the subject's world in the process.

Animals, which have either undergone natural or recreative invasion by the agent, would partly take within them the effect of the invasion on the observed behavior of the subject. A primatologist who captures a chimpanzee and transports the animal to the laboratory in order to observe it, has infused the effect of the agent on the world of the animal. The same would be true to the physical anthropologist who goes to the jungle and studies the gorillas in the wild. His presence in the field of observation is an invasion on the world of the gorillas and some of their behavior though exhibited in the animals' own habitat, would carry some of the effects or interactions of the agent. This would bring in some reconstructive elements of the agent's invasion.

Now things would be a little different on objects with the least amount of reconstructivity that the invasion of the agent with his objective to reconstruct the observed would be largely dependent on the manipulation and control of the agent. An example is a chemist who works on molecules of certain compounds in order to produce certain chemicals. The chemical process changes the composition of the compound according to the desired outcome which the chemist intends.

This is also the same with a geneticist who has determined the phenotype of a certain gene and shoots it with another strand of gene from another organism in order to manipulate its characteristics. Yet all these are invasions, unless the compounds are unstable that the molecules recreate themselves in the advent of some environmental conditions or that the DNA of the gene has mutated that its phenotype has inherently changed.

But this would not exactly be the same with the atomic particles. The atomic particles possess a certain degree of uncertainty as to their location and velocity. Then probabilities would result had we performed different invasions in order to observe it. Heisenberg found out that bombarding the electron with electromagnetic waves in order to ascertain its velocity and location results in uncertainty. This is reconstruction as a result of invasion. But by positing that this is really the inherent nature of atomic particles is also the same as saying that the reconstructive potential is inherently possessed by it. If invasion makes the atomic particles behave in a crazy manner, Quantum Theory suggests that it is inherently crazy by nature. This being the case, then shooting a particle with another accelerated particle at high energy would create possibilities. The process is like smashing a stone with a sledgehammer in order to know what it is made of. But smashing a particle with another particle may break it up into constituent parts with its trails ascertained to determine what it is. But since uncertainty governs the subatomic world, then with the high-energy bombardment uncertainty or probability would also suggest that the debris in the smashing may be reconstructed high-energy wave-particle that assume certain mass due to high energy and high frequency. While they may really be constituent particles, possibilities would also suggest that they are reconstructed or recreated particle-waves resulting from high-energy invasion. And if string theory suggests that particles are like vibrating strings, then these strings may have been violently strummed that it produced vibrations at higher frequencies producing higher energy and evidently manifesting itself with mass. The whole process then is the result of invasion that science cannot do away with. And with the reconstructive character of the observed, reconstruction also has taken place.

At the quantum level, the observer becomes part of the observed. The greater that the observer disturbs the observed, the more that he may become part of it. As more energy is dispensed to a particle to smash another, how sure are we then, that the traces of light that are produced are images of constitutive particles of the smashed or are they particles newly created because of the great amount of energy that the observer has used? "God does not only play dice but the greater the smash, it seems humans have played God."

With theories then as means to textualize the world, this project assumes that theories are able to reflect these realities. That being the case, then how is this reflection bridged? How does this reality at the end of the pole reflected at the other end of it?

Between Theory and the World

What bridges the gap is interpretation. Measurement is quantified interpretation. Meter, inch, foot, miles, yards, etc. are quantitative interpretations of length, breadth and depth. Gram, ounce, pound are quantified interpretations of mass and weight. Fahrenheit, Centigrade, Kelvin are quantitative interpretations of temperature. Seconds, hours, minutes are quantified interpretations of time or duration. Combinations of these units would characterize speed (meter/second), force (kilogram-meter/second^2) energy (kilogram-meter^2/second^2) etc. There are just some constants in the world. The speed of light and Planck's quantum number are just a few of these constants. These constants, however, are inherent in the milieu-already-constructed. Other than these constant quantities, the world is also built with geometrical shapes. These shapes have sizes. And these shapes and sizes can be measured. Such measurements can be standardized in the form of numbers. Numbers, however, are characterized with precision and independence. The quantity 1 is the same to represent one unit and not any other quantity. The same with other quantities. Though how big or small the object being measured, numbers could always give an account of it. Thus to the smallest function or to the highest scale, as long as there are zeros to attach to these numbers, the object can be measured. Gram, meter, second, etc. are units of interpretation inasmuch as measurement is quantified interpretation. Since numbers are a representation of measured interpretation, amazingly however, numbers can also be used to represent relations that have their own definite operations. These operations have defined rules which when substituted with certain quantities could represent realties.

Mathematical operations that represent the function of "X" and "Y" are governed by rules that interpret these relations. Mathematics therefore is rule-governed interpretation of relation among quantities and their operation. Mathematics that abstracts these relations and operations could stand on its own. It is characterized with independence and self-containment. This means that the rules that underlie the operations of numbers create an existing reality of its own. This is axiomatic reality as previously discussed. Thus numbers using these rules can be factored, multiplied, added, subtracted, divided, substituted even if they don't signify anything but are just abstractions at the moment. It is true that:

$$a^4 b^{-3} c^5 / d^{-1} e^2 f^3 \quad \text{is always equal to} \quad a^4 c^5 df^3 / b^3 e^2$$

This is true as long as a,b,c,d,e,f are not equal to zero. Anything other than this result would therefore be untrue. We have arrived at the solution to this pure abstraction though we do not know what a,b,c,d,e,f stand for. But the procedure and the solution are always true for the rules that lie beneath the operation even without representation to observable objects exist. Thus 1 added to 1 is always 2 because the law of addition permits the operation. In the same manner,

$$((x^2 + y)^{1/3})^3 = x^2 + y$$

This is because the law of radical says,

$$((a)^{1/n})^n = a$$

If numbers are independent and could exist on their own, they could be used to stand for something in the world. Let's say 5 tomatoes which can be described through their weight of 250 grams. If we take 5 more tomatoes weighing at 200 grams then the total number of tomatoes would be 10 with a combined weight of 450 grams. Measurement then, describes the tomatoes as objects in the world, the numbers represent them and the act of combining the objects with their representations is described by mathematics. For Kant, this is knowledge *a priori* or knowledge obtained through pure reason and not through sense-impression (Stewart and Blocker 1996: 238).

We now have these forms of interpretation to describe and represent the world. But not all acts or facts can be quantified to be interpreted by mathematics. Inasmuch as mathematics uses the rules of logic, logic itself is interpretation of the states or conditions of things, persons, or events using rules of inference. If mathematics utilizes quantities and symbols, logic uses language as its currency for operation, although symbolic logic uses symbols to transform these states and conditions into representations. The rules of inferences, just like mathematics, are also independent and could stand on their own. Thus if we want to logically deduce the condition of the weather by inferring on some given states of affairs, we could say:

If the sky is heavy of nimbus clouds and it is humid,
    then it will rain.

But the sky today is filled with dark nimbus clouds and it is humid,
    therefore it will rain.

In another occasion,

It is not raining,
    therefore the sky is not heavy of nimbus clouds and it is not humid.

The independence of inferential rules in logic creates its own axiomatic reality. All these forms of interpretation (measurement, numbers, mathematics and logic) are characterized with independence and self-containment, with which they create realities of their own. These forms of interpretation then can bridge the gap between theories as texts and the world which it seeks to textualize, making theories an interpretation of the world itself. Measurement, numbers, mathematics and logic are interpretations fit for the milieu-already-constructed and some aspects of the milieu-under-construction. But the milieu-under-construction in the area of social sciences has theories which account for conditions that are not measurable. With this the gap that bridges theory and the world are interpretative or creative statements arrived at through reason. These interpretative statements are creatively crafted to qualitatively penetrate and give an analytical account of the social phenomenon evolving (Figure 6.1). But as described in Chapter 4, this system of statements has the capability to reconstruct the world of the subject, even the observer himself. As the theory-statements, therefore, provides qualitative description and analysis of the phenomenon which patches the gap between the theory and the world it seeks to textualize, the world of the subject or the observer is reconstructed in turn.

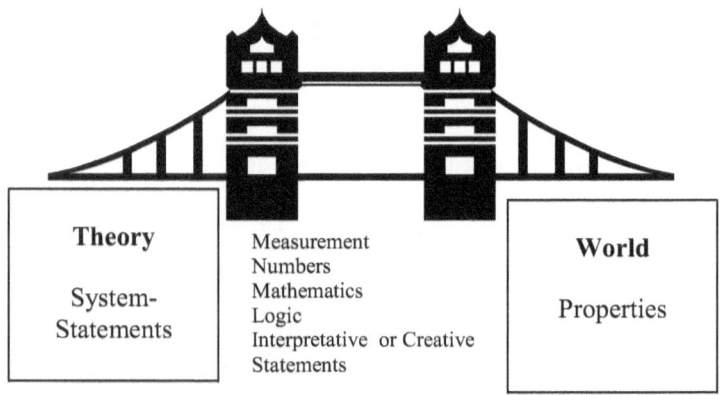

Figure 6.1 Bridge between Theory and the World

The next question then would be, how faithful is the theory in its representation of the world? Different schools of thought would provide different answers to this question. For the realists and empiricists, experience and its correspondence with the prescriptions of the theory would determine the theory's faithfulness. For the constructivists, its confrontation with other theories and its capacity to account for anomalies which another theory could not explain despite the regularities they could both account for, would make another theory more faithful in its representation of the world.

The problem here lies on the fact that we don't have a measuring stick to calculate the faithfulness of theories.

Measurement, numbers, mathematics and logic are themselves forms of interpretation just like theory. Theories which account for patterns of regularities are also confronted with irregularities which they could not give account of. Thus with theories consisting of system-statements and the world on the other, these theory-statements are characterized with relations which can be translated to mathematical equations. These symbolic equations can be substituted with numbers arrived at through measurements or if the aspects of the world are qualitatively constituted, the interpretative and creative character of these statements could give account of the regularities which the theory-statements could consistently textualize. Outside of these conditions are irregularities that the theory-statements could not give account of. If the amount of regularities that the theory-statements could interpret would be the measure of the theory's faithfulness, then it breeds another big problem. Theories that have successfully given account of irregularities which lie outside the explanatory power of another theory exposes another set of anomalies. It is like patching a crack on a piece of pottery. As you apply some filler, you'll find more cracks as you sand and clean it. The amount of regularities that a theory could give account of the world compared to another is, therefore, not the measuring stick. When Kepler formulated the Law of Planetary Motion, which posited the elliptical orbit of planets, an irregularity arose on account of gravity that the law was silent about. Gravity was successfully explained by Newton's Universal Law of Gravitation, postulating that it is a pull that the body exerts on another as it orbits around the body. But such explanation exposes an anomaly of why Mercury's orbit behaves eccentrically that it recedes $0.01°$ in a century as it moves around the sun (Ford 1974: 1099). Einstein, however solved the puzzle only to expose another anomaly about the nature of light that bends on the path of warped space-time. But how about observation? Can the number of observables that the theory could interpret qualify for a better theory than the other? In the same way, can the quantity of observables that the theory could not give account of make it an inferior theory?

We have to be reminded that that there are theories in science that have already been abandoned. One of which is the theory of ether. It was suggested that light travels on a substance called ether. But the Michelson-Morley experiment debunked the idea. Their experiment in 1887 where they shot rays of light, one ray at right angle to the earth's spin and one parallel to it, did not show any discrepancy as to their reflection, permitting them to conclude that ether does not exist to influence the speed at which light travels. The idea of ether was completely thrashed when Einstein proposed his special theory of relativity that light travels at constant speed even if the source is moving. The concept of ether then was completely unnecessary (Gamow 1961: 93-98).

But in view of the observables, whether they could be used as criterion to ascertain a better or inferior theory, we have to be bear in mind that observables by themselves mean nothing. Observables by themselves don't make sense. Observables need to be organized. They need to cohere to some form of principle or idea before they can be used. Moreover, these observables need to be interpreted. Thus it would take another form of interpretative device, a theory or principle, where these observables could cohere if they were to confront another theory. In the same manner, if no theory exists yet to permit the organization of these observables, then at least an emerging principle stands at the back of the scientist or theorist's mind that compels him to assess the theory at hand. It is not, therefore, the observables that would hope to stamp the verdict of which is a better or inferior theory but the idea or the emerging principle that would provide coherence to these observables. But will it be enough? Remember, one principle might bring about another irregularity.

The Milieu-of-the-text

Is there any hope, however, of determining how one theory is better or more faithful than the other? For Kuhn, which will be discussed in the succeeding section, "there is none" for it is a matter of one scientific communities' conviction to a paradigm but for Popper "there is a method of appraisal through falsification." This being the case, then we are left with only one choice, either we have a hoax or we have something that works. Either we have a useless idea or we have a theory that does its job. We could partly thank the theorist or scientist to whom we owe a lot of creativity. The point is, the theory works. The text which consists of theory-statements could give account of some aspects of the world and be an interpretative device to express this account. The point is, if we seek to confirm the theory's faithfulness by doing the experiment, what we did was to manipulate, disturb and replicate the same manipulation and disturbance that the theory was able to incorporate in its textualization. Thus the theory-statements contain textualized disturbances that the faithfulness of the theory would also be difficult to determine. Whether we have a hoax or a thing that works would be difficult to ascertain for no measuring stick is available. Is this a cause for disenchantment? Far from it. This is a cause for relief, if not jubilation, for the more that theories expose anomalies and the more that we search for the yardstick to ascertain the faithfulness of theories, the more that we witness science work because the more that humans invent and the more products of inventions humans have produced.

Yet if we insist on a yardstick, we let Kuhn and Popper debate each other. But before moving into the debate of the two greats in the philosophy of science, let's revisit the two greats in physics, whose works have been the main dish among philosophers of science to cite.

Newton vs. Einstein

Story or myth has it that Newton sat under an apple tree and, in his deep moment of contemplation, an apple fell. Pondering on such an occurrence and building on Galileo's constant acceleration of free-falling bodies, he formulated his Universal Theory of Gravitation. But if Einstein were sitting under the apple tree and an apple fell, he would have picked it up and stared at the indention on the top of the apple around the stem. That, he would exclaim, illustrates gravity and explains the General Theory of Relativity.

For the apple that fell and any object for that matter, Newton's concept of gravity says that of a force that pulls the object towards the earth. Newton's Law of Motion can be summarized into the concept of motion and force that causes an object to move. Newton's laws explain how an object moves from one direction to another with the employment of force and how it would behave if it confronts another object in motion. Newton's next project then was gravity. Remember, force is the push or pull exerted on an object in order to cause it to move. Anything when thrown upward falls down. Such movement when examined in a bigger scale would make the earth look like a massive ball and the body that falls on it would be considered a body attracted towards it. Newton's predecessor, Galileo had previously examined the phenomenon of falling bodies when he concluded that free-falling bodies no matter the mass have the same acceleration equal to 9.8 m/sec^2. Newton later generalized the phenomenon to propose his Universal Law of Gravitation on the presumption that gravity is force. It is the same force that keeps planets in their orbit around another body.

$$F = G(mM/r^2)$$

The Milieu-of-the-text

The force (F) is dependent on the mass of the two bodies (mM). The greater the mass the greater the attraction but the greater the square of the radius (r^2) of the two bodies, the lesser the force. G is the gravitational constant computed at 6.67×10^{-11} m^3/kg-sec^2 which accounts for the generalization of such motion and constancy of acceleration. Thus if you throw a body strong enough to make it escape the threshold of attraction, the body would orbit around it. Move farther, the force decreases and the object flies away from the body.

Gravity therefore is conceived to be an attractive force of a body acting on another body orbiting around it. The model would be like a ball tied to a string which you swing around your arm. The acceleration and force exerted on the ball moving around your arm would be calculated in this manner.

$$a = v/t$$
$$a_{circ} = v^2/r$$
$$a_{circ} = (2\pi r/t)^2/r \text{ or } 4\pi^2 r/t^2$$

The distance of the two bodies is denoted by the length of the string (r) which is actually the radius of the circle while $2\pi r$ is the circumference of the circle which pertains to the distance the ball travels around your arm at a certain duration (t). The force exerted on the ball to keep it swirling around your arm while preventing it to move away would be calculated by multiplying acceleration with the mass (m) of the ball.

$$F = m((2\pi r/t)^2/r) \text{ or } m(4\pi^2 r/t^2)$$

This could then be combined with the formula describing Newton's concept of gravity.

$$G(mM/r^2) = m(4\pi^2 r / t^2)$$
$$GmM/r^2 = m4\pi^2 r / t^2$$
$$GMt^2 = 4\pi^2 r^3$$

Through this formula, the mass (M) of an object at the center could be computed given the orbital period (t) and the distance (r) of the body at the center. This also confirms Johannes Kepler's Third Law of Planetary Motion which says that the ratio of the cube of the average radius of orbit of a planet to the square of the time is constant for all bodies rotating around a body in space. An example of this is our own solar system where the ratio of the cube of the radius of orbit and the orbital duration is 2.51×10^{13} m^3/t^2 for all planets orbiting around the sun.

$$GM / 4\pi^2 = r^3 / t^2$$
$$K = r^3 / t^2$$

Kepler's Law of Planetary Motion explains the behavior of planets around the sun. Newton, however, provided the reason why. But Newton's theory is founded on the idea that the reference frame from which the body is being measured is presumed to be stationary. A person by the roadside watching an automobile pass by does not find the road moving with it. It is only the automobile moving and not the road. Moreover, the formula for characterizing speed presumes no threshold. If a body is 200,000 miles away and a spaceship reaches it in a second, the speed of the spacecraft would be 200,000 miles/sec. Would this be permissible? Newton's formulation is silent on this.

It has been discussed how Einstein united space and time in his Special Theory of Relativity. Einstein felt this was incomplete. The next phase of his project was to generalize it to incorporate gravity. Einstein was not satisfied with how Newton's Universal Law of Gravitation accounts for the behavior of bodies in space. Newton's gravity is basically

The Milieu-of-the-text

founded on the concept of attractive force. The model of one body orbiting another body in space would look like this. Cut a piece of string and tie one end on a ball. Hold the other end and swirl it around you. Newton's Law of Gravitation can be likened to the string. The attractive force that holds two bodies in space is like the string that keeps the ball in orbit around you. But what if we let go off the string. The ball will fly away from us at the same time. That picture though simple has grave consequences to reality and theory.

What if the gravitational pull of the sun on the earth was suddenly cut off? That would result in the earth instantaneously flying out of its orbit. The earth is approximately 98 million miles from the sun. If the sun disappears in a second then the earth flies in a second as well. Thus if we consider the speed of light in this regard, Newton's Law of Gravitation would be found wanting. It could not account for speed of light which is constant. With this irregularity, Einstein's next project was to incorporate the speed of light in the configuration of planetary motion and, in turn, formulate the General Theory of Relativity.

With this, his attempt was first to challenge the idea of action at a distance which is implied in Newton's theory of gravitation that gravity is a force acting on two bodies in space. The greater the mass and the lesser the square of distance separating the two bodies, the greater the force between them. If this is the case, how come when we place a ball inside the box and suddenly raise it, the ball stays on the floor of the box inside. Lower the box and the ball sticks on the ceiling of the box. Move the box to the right and the ball goes to the left. Swing the box to the left, the ball goes to the right wall of the box. While Newton would call this the action-reaction phenomenon, Einstein calls this the equivalence principle. Gravity is nothing but the effects of an accelerated system. If gravity for Newton is action at a distance, gravity for Einstein is an accelerated system. Since relativity states that everything is in motion, then in a local frame, the physical effects brought about by gravity is the same as the effects created by accelerated motion. The next question then would be how could the structure of bodies in such an accelerated system be configured? If for Newton it is like a ball attached to a string, how could general relativity create a picture of it?

Here Einstein's genius emerges again. Rather than a force between two bodies in space, gravitational field is the geometry of space-time. It is the indention or warp in space-time brought about by mass-energy on it. Stand on your bed. You will see that the bed indents towards you. For general relativity, this is gravitation. Without mass-energy, there would be no indention, there would be no gravity. Thus a lesser body in space would be caught in that dent. Instead of being secured by a string, the lesser body keeps its orbit because of the dent created by the massive body. The lesser body seeks its shortest distance like a straight line, well in fact it is moving around the massive body because of the dent. Its movement around the body though it assumes it is just a straight line is its geodesic. General relativity will state that mass-energy creates a curvature in space-time. Einstein already had a picture in mind. But something was lacking. He didn't have the equation for it. With this principle, Einstein took almost 10 years to come up with the proper mathematical expression. But it did not come without the previous findings of a German mathematician, Beinhard Reimann years before him. From Reimann, Einstein found the holy grail of his equation. Reimann introduced the geometry of curved spaces which challenged the purely Euclidian geometry of flat spaces. The test for a flat surface would result in a triangle having a total angle of 180° based on Euclidian geometry but Reimannian geometry would find the same triangle having lesser or greater angular measurement than 180°.

If you draw a circle and dissect it into 4 parts each quarter of circle near the center would measure 90°.

The Milieu-of-the-text

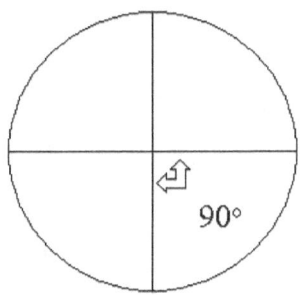

Figure 6.2 Flat Circle with 90° Component

But half a sphere whose base is also circular when divided into 4 parts would have more than 90° measurement for each part.

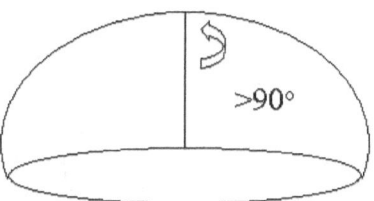

Figure 6.3 Sphere with >90° Component

On the other hand, a trumpet-shaped curvature when a quarter is taken would measure less than 90° inside.

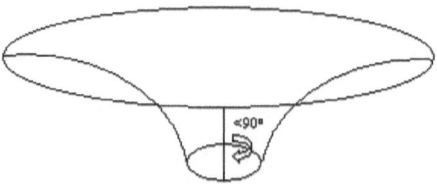

Figure 6.4 Trumpet Figure with <90° Component

For Euclidian geometry, the shortest distance between two points is a straight 180° line. Remember that the sum of all the angles of a triangle is also a 180° line. But for Reimannian geometry, the shortest distance between two points is a curved line greater or less than 180°.

An event that happens in a Euclidian space occurs between two points.

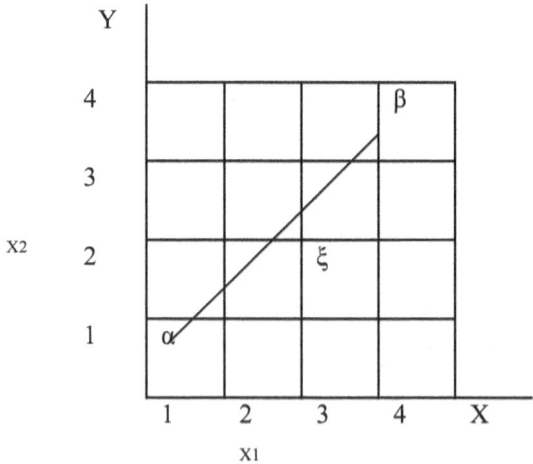

Figure 6.5 An Event in Euclidian Representation

Using Pythagorean theorem, the distance between points α (alpha) and β (beta) of event ξ (xi) can be measured by the formula:

$$\xi^2 = (X^1_\beta - X^1_\alpha)^2 + (X^2_\beta - X^2_\alpha)^2$$

If the dimension were increased to three, the expression of such distance would be found in this coordinate system:

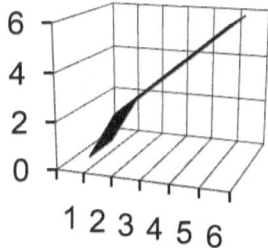

Figure 6.6. Three-Dimensional Coordinate System

$$\xi^2 = (X^1_\beta - X^1_\alpha)^2 + (X^2_\beta - X^2_\alpha)^2 + (X^3_\beta - X^3_\alpha)^2$$

Euclidian geometry adopts such dimensions that run to three. Beyond that, Euclidian geometry stops. It is understandable for Euclidian geometry is applicable only on flat surfaces. There will only be three dimensions which could be drawn in the form of a cube. Since the cube is complete, an addition of another side will be ridiculous to account for the fourth dimension. There is no room for another side to be added. The fourth dimension then is time.

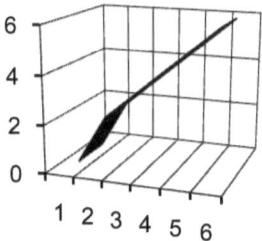

Figure 6.7 Three Dimensional Space Plus Time

In order to extend this summation, a theoretical imagination has to work, incorporating time x^0 as another dimension. In this case, an event that happened in four dimensions can be calculated using the following representation:

$$\xi^2 = -(X^0_\beta - X^0_\alpha)^2 + (X^1_\beta - X^1_\alpha)^2 + (X^2_\beta - X^2_\alpha)^2 + (X^3_\beta - X^3_\alpha)^2$$

This would mean that in space-time, event ξ occurred between points α and β, at x^0 (time) and spaces x^1, x^2, x^3. But even with the incorporation of the fourth dimension, Einstein's theory of gravitational field is not yet fully expressed.

The Milieu-of-the-text

Einstein's theory of gravity, better known as the General Theory of Relativity states that mass-energy content produces curvature in space-time.

At this point, Reimann's tensor analysis of curved spaces work. Reimann's tensor seeks to describe the curvature of space at every significant point. With a curved space-time, Reimann's manifold system is designated by a metric tensor "g" which would describe an event that happened in a coordinate allowing two input units u and v or v and u for each tensor g. It is equivalent to g(u.v) or $g_{\alpha\beta}$.

$$\xi^2 = -(X^0_\beta - X^0_\alpha)^2 + (X^1_\beta - X^1_\alpha)^2 + (X^2_\beta - X^2_\alpha)^2 + (X^3_\beta - X^3_\alpha)^2$$

$$\xi^2 = -(X^0_{\beta\alpha})^2 + (X^1_{\beta\alpha})^2 + (X^2_{\beta\alpha})^2 + (X^3_{\beta\alpha})^2$$

$g_{\alpha\beta} = \alpha$

		β = 0	1	2	3
	0	g_{00}	g_{01}	g_{02}	g_{03}
	1	g_{10}	g_{11}	g_{12}	g_{13}
	2	g_{20}	g_{21}	g_{22}	g_{23}
	3	g_{30}	g_{31}	g_{32}	g_{33}

Figure 6.8 Reimann's Metric Tensor

Since $g_{\alpha\beta}$ is the same as $g_{\beta\alpha}$ then there will only be 10 significant tensors instead of 16 as presented in the matrix (Figure 6.8).

$g_{\alpha\beta} = g_{\beta\alpha}$

g01 = g10; g02 = g20; g03 = g30

g12 = g21; g13 = 31; g23 = g32

All the significant tensors would be:

g00, g11, g22, g33, g01, g02, g03, g12, g13, g23

They could be summarized into:

$$ds^2 = g_{uv}dx^u dx^v$$

It would then be presumed that an empty space-time would yield zero curvature. This is a flat space where Pythagorean geometry works while Riemannian geometry yields zero.

$$R_{uv} = 0$$

But a space with mass-energy content would create a depression with tensors not equal to zero.

$$R_{uv} < 0$$

Reimann's geometry was actually formulated even before Einstein worked on his project of generalizing relativity. It was in 1900 when Einstein birthed the Special Theory of Relativity. Then years after, he became successful at generalizing it. But it was in 1854, sixty years ahead of him when Reimann embarked on his lecture on tensor analysis, and 60 years hence when such geometry of curved space found meaningful theoretical home. What Einstein did was to exactly fit in Reimann's metric tensor to describe the curvature of space in order to mathematically express the concept that curved space-time, which can now be understood as gravitational field is warped by the presence of mass-energy on it. Einstein devised a formula to express this principle (Kaku 1994:92, 98-99 see also notes).

The Milieu-of-the-text

$$R_{uv} - 1/2 g_{uv} R^{\alpha}{}_{\alpha} = (-8\pi G/c^4) T_{uv}$$

Reimann's curved space is on the left side of the equation which is equal to the energy-mass content of space-time expressed by the tensor summation T_{uv} multiplied by the gravitational constant G and 8π over the speed of light. Gravity therefore is not simply force that attracts bodies at a distance apart from each other, but gravity is the geometry of space-time described by being warped or dented (Figure 6.9).

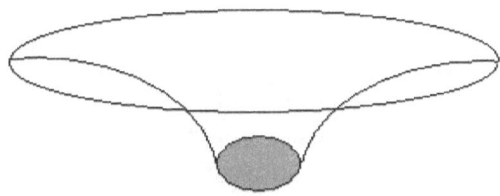

Figure 6.9 Warped Space-time

In this case, the General Theory of Relativity now solves the problem posed earlier. If the sun suddenly disappears, Newton's universal law of gravitation would say that earth would suddenly fly out like a ball cut out of a circling string. But with the picture of a geometrized space-time, earth will fly out approximately 8 seconds after, the same time light from the sun arrives on the earth.

Let's go back and revisit the apple. Newton noticed the apple fall on the ground. He theorized that the attractive pull of the earth caused the apple to fall. That is the Universal Law of Gravitation at work. But what if Newton fell from the tree together with the apple beside him? Well, he would have noticed the apple to be at rest because both of them were falling at the same time. That is the Special Theory of Relativity accounting for the occurrence. Then with the apple, Einstein could have picked it up and looked at the indention on top. That is the General Theory of Relativity illustrating gravitational field.

Did Newton make a mistake? Was his theory false? When Einstein's theory was accepted Newton's theory was not abandoned. Its calculation is still being used to ascertain the mass of two bodies experiencing gravity in space. In order to make a sense of the seeming rivalry, philosophers of science mapped out the development of these ideas.

The Rise and Fall of Enlightenment

Human fascination with the heavens from the ancient times originates from the observation that the survival of humans is dependent upon the condition of the heavens. The patterns and positions of stars would suggest the seasons. The seasons will indicate when to plant and when to harvest. This would signify that the heavens govern the routines of life on earth. The ancients plotted the path of planets and even drew the patterns of stars. It is with this thought that philosophers assumed that the nature of things down here is dependent on the things up there, elevating the idea all the way towards finding out how the heavens operate. Aristotle geo-centricized the known universe while Ptolemy provided the mathematics of its nature. It was this fascination that extended to the enlightenment period where Aristotle's concept of the universe was challenged. Copernicus helio-centricized the universe, Kepler provided the mathematics of the planets' motion, while Galileo used the telescope to prove Copernicus was right as Newton universalized the motions into law.

But the period of Enlightenment was not only about nature and the physical world. The use of reason was not all meant to discover the laws of nature but also to find out the best ordering of society and serve humanity for its good. Reason will lead to freedom and freedom will lead to happiness. Happiness as the end of society is the classic Aristotlean revival. While the English Enlightenment found Newton's lead to discern the laws of nature, the French looked up to Jean Jacques Rousseau to use the social contract in organizing the society with the individual experiencing the liberty he is presumed to enjoy. The Germans came late in swifting through the current of Enlightenment but found Immanuel Kant to be its champion.

Kant was embroiled with the intellectual feud of his time between Locke's empiricism when he argued that knowledge originates from sense impression on the one hand, against Descartes' rationalism when he advanced the existence of innate knowledge on the other. Kant proposed the middle ground when he published the *Critique of Pure Reason* in 1781. The rise of modern physics, which threatened morality and the existence of God, was the offshoot of the Enlightenment's crisis. God does not figure in a science where reason is its main instrument of discovery and organization. This is Kant's inspiration in writing his book while re-situating metaphysics as a method of advancing knowledge. Kant admitted that knowledge can be derived from experiences of the senses (*a posteriori*) but also claimed that there are general laws like that of mathematics which can be deduced from rationally uncovered laws (*a priori*). Reason is the source of human understanding of general laws and the means to understand God's existence and moral precepts (Rohlf 2010).

But the Enlightenment which could be reckoned to have begun in 1700 to deify reason as the means to discover the physical laws and organize society met its downfall in 1800 after the French Revolution in 1789. The French preoccupation to find the greatest good using reason ended up in violence at the aftermath of the revolution resulting in excesses and authoritarianism. The French reverted to a more parochial and simple philosophy of romanticism emphasizing subjectivity and personal emotions to locate a person's worth rather than the scientific objectivity of the Enlightenment period. The author of romanticism was no other than Rousseau himself.

Interlude with Modernity

The demise of the Enlightenment, however, did not lead to the total abandonment of reason but the emphasis on personal emotions and subjectivity brought with it the preeminence of the self rather than the whole society. With the death of the Enlightenment project came two roads to its mourning. One road led to modernity - an epoch without defined historical demarcation after the Enlightenment – but more of an attitude marked with a) increased rationalization of social systems but b) the increased desperation of the individuated self. Modernity was born at a time when Europe was catapulted into the Industrial Age when factories were being established which brought about high specialization in tasks to efficiently produce a desired material result. Factories brought about a different sociological environment where supposedly rational beings are warehoused in a factor to accomplish a collective task. But factories that house the mechanized means of production would not have been established with the use of capital, thus capitalism became a heightened doctrine. But it was also at this phase where Marxism was the dialectic doctrinal alternative since the ascension of capitalism to its height brought with it also the oppression of the workers.

It would be remembered that the Enlightenment project was not just a revolt against the impositions of the Catholic Church on things that border on knowledge but it was the enshrining of reason to discern the physical world and reorder our social lives. But the re-ordering of social lives should come with the recapturing of Aristotle's end of bringing happiness to our lives. Reason should bring about the fruits of freedom and happiness. Rousseau's project of social contract puts this as:

> "The problem is to find a form of association which will defend and protect with the whole common force the person and goods of each associate, and in which each, while uniting himself with all, may still obey himself alone, and remain as free as before" (Rosseau 1981:18).

How can one feel free after he or she has given part of his or her freedom for the re-ordering of his or her social life in society is an irony in itself. Reason would build an ordered society. Rationality was aimed at constructing a well-functioning, well-ordered system. The end result of these well-functioning systems which reason has built should be happiness to the individual selves consisting these systems.

The rationalization thesis that defines modernity was the handiwork of the German sociologist Max Weber with the publication of his seminal work *The Protestant Ethic and the Spirit of Capitalism* in 1905-1906. In his book, he found out that the values of frugality, fair play and hardwork that defined the Calvinists of his time were the very factors that contributed to their prosperity and propagated the capitalistic society. These values could wax off its Protestant color but the values remained which maintained the capitalistic system. This was what he called rationalization. It is the mechanistic conversion of values and even personal emotions into rational calculations. The development of capitalistic society also came in parallel with the development of a bureaucracy which is characterized with the hierarchical ordering of positions, specificity of tasks, objectivity of operation and documentation of communication all fall in to rationalization that make up for an efficient system. The idea of control holds the social system together and it is this mechanistic conversion of values that make up for bureaucratized society and also the same rationalized and mechanized values that make the capitalistic system to work. Control is essential and inevitable. Thus an individual, in order to make for the rationalized system would have to inevitably give himself or herself up to the system creating what Weber referred to as *stahlhartes gehäuse* (housing hard as steel) which Talcott Parsons translated as "Iron Cage" (Cole 2015). The essence of Weber's thesis is that, capitalism had developed a bureaucratic society which was brought about by rationalization. But the more we make up for a rationalized or bureaucratized society in order to achieve prosperity, the more that we become mechanized, objectified or even "iron caged" that we become depersonalized and we lose the sense of self. The more that humans create a rational society, the very essence of recapturing freedom is lost for one would feel more enslaved because he or she wants to make up for an efficient, rational system.

> The bureaucratic organization, with its specialization of trained skills, its delineation of competencies, its rules and hierarchical relations of obedience... is ... in the process of erecting a cage of bondage which persons – lacking all powers of resistance – will perhaps one day be forced to inhabit, as the fellahs of ancient Egypt. This might happen if a purely technical value – a rational civil service administration and distribution of welfare benefits – becomes viewed as the ultimate and single value in reference to which the organization of all affairs ought to be decided. The bureaucracy achieves this result much better than any other structure of domination (Weber 1968:1402 as quoted in Kalberg 2001:179).

Weber argued that it is rationalization that alienated the individual in a capitalistic society rather than the domination of the bourgeoisie who owned the means of production and who had created the political superstructure as Marx posited. Marx also contended that the fetters that the proletariat suffered were due to the consequences of domination brought naturally by the capitalistic order. Weber countered this with his position that it is the *stahlhartes gehäuse* that had become the means of domination which alienated the individual from the freedom he or she ought to have.

Emile Durkheim a French sociologist also had a similar analysis of the rise of industrial societies in Europe. With the publication of his book *On the Division of Social Labor* in 1893, he positioned that the division of labor was the defining feature of the industrialized societies in Europe that had factored the element of increased individuated self. He argued that the specificity of tasks, the mechanized motions of the human body in a factory has brought the autonomy of the individual. The more specialized the task, the more autonomous that the individual should efficiently do the work. The more the demands that society ought to accomplish the more division of work the society should have, the lesser the collective experiences. This brings about the individuated self which is the defining characteristic of modernity (Carls 2016).

But there is a dark side to this. The depersonalized and autonomous self which habitate in a highly capitalistic society experiences alienation and desperation. The collective feeling of frustration despite the individuated self experiencing freedom and material things in a capitalistic society is a phenomenon he called anomie. He identified this feeling of despair and disaffection on the division of labor and rapid social change as embodied in his book On *the Division of Social Labor* in 1893.

Addressing the social pathologies of modernity was the objective of the *Institut fur Sozialforshung* or the Institute for Social Research otherwise known as the Frankfurt School.

Established in 1923 with the money of a grain exporter Felix Weil, the institute was also known for its Marxist studies and the leading proponent of the critical theory in sociology, a theory that was meant not just to explain but also to change society. Among the prominent figures of the first generation of critical theorists were Max Horkheimer, Theodor Adorno, Herbert Marcuse, Walter Benjamin, Friedrich Pollock, Leo Lowenthal, Eric Fromm and Jürgen Habermas. Critical theory took its first strides from Marx himself with his assumptions that the working class will take over the means of production as a result of the oppression they suffer from the capitalists and demolish the system in favor of one which is a socialist. Being Marxist in its leaning, the Frankfurt Institute was meant to take its aim at the critiques of modernity with one specific inquiry as to why humans had not tried to liberate themselves despite the domination they suffer in a capitalistic society. Marx took its blame on the capitalist order but the coming in of different disciplines including psychoanalysis led the institute to divert its project to the issue of emancipating the self. With the rise of Nazism in Germany, Hitler aimed his guns at blaming the Marxists and the Jews for their defeat in World War I and the loss of their territories. The Frankfurt scholars migrated to the United States where they found their home at Columbia University in New York (Corradeti 2016). Among these scholars who found refuge in the United States in 1933 was Eric Fromm. In his book *Escape from Freedom* published in 1941, Fromm summed up his thesis in the following way:

> So far this book has dealt with one aspect of freedom: the powerlessness and insecurity of the isolated individual in modern society who has become free from all bonds that once gave meaning and security to life. We have seen that the individual cannot bear this isolation; as an isolated being he is utterly helpless in comparison with the world outside and therefore deeply afraid of it; and

because of his isolation, the unity of the world has broken down for him and he has lost any point of orientation. He is therefore overcome by doubts concerning himself, the meaning of life, and eventually any principle according to which he can direct his actions. Both helplessness and doubt paralyze life, and in order to live man tries to escape from freedom, negative freedom. He is driven into new bondage. This bondage is different from the primary bonds, from which, though dominated by authorities or the social group, he was not entirely separated. The escape does not restore his lost security, but only helps him to forget his self as a separate entity. He finds new and fragile security at the expense of sacrificing the integrity of his individual self. He chooses to lose his self since he cannot bear to be alone. Thus freedom – as freedom from – leads into new bondage (Fromm 1965: 282-283).

Fromm argued that freedom is an innate quality of humans and if he or she cannot live with it, he or she will turn to authoritarianism or fascism. But living with freedom under a democracy otherwise reduces the individual into an automaton which is a way of escape.

This particular mechanism is the solution that majority of normal individuals find in modern society. To put it briefly, the individual ceases to be himself; he adopts entirely the kind of personality offered to him by cultural patterns; and he therefore becomes exactly as all others are ad as they expect him to be. The discrepancy between "I" and the world disappears and with it the conscious fear of aloneness and powerlessness. This mechanism can be compared with the protective coloring some animals assume. They look so similar to their surroundings that they are hardly distinguishable from them. The person who gives up his individual self and becomes and automaton, identical with millions of other automatons around him, need not feel alone and anxious any more. But the price he pays, however, is high; it is the loss of his self (Fromm 1965: 208-209).

Conformity becomes a way of escape in a modern democratic society. The fear of aloneness, the lack of spontaneity, though vested with all the freedom to express oneself becomes a societal ill.

The automatization of the individual in modern society has increased the helplessness and insecurity of the average individual. Thus, he is ready to submit to new authorities which offer him security and relief from doubt (Fromm 1965: 230).

Freedom that should be enjoyed in a modern society leads to another defect. The more rationale modernity has built the society, the more alienation the individual would have felt. The more bureaucratized and labor specific the rationalized social system has become, the more individuated the self as a consequence; the less freedom the self experiences and the more despair the self suffers. Reason should have liberated the self but the more it has become iron caged only for the sake of creating a rational and efficient system. Modernity has invested so much premium on making social systems rationale only to enslave the self and instead of bringing happiness to the self as the fruit of reason, the more that it had brought despair on the individual. Humans have built a rationalized system but they have ended up with an irrationalized self.

Jürgen Habermas, the next generation of critical theorists was first connected with the Frankfurt School but his intellectual differences with the other figures in the institute led him to separate ways with them. If the Frankfurt School's stand was addressing the emancipation of the self, Habermas assumes that the mediating factor between the rationalization of the social system and the emancipation of the self is communication. He introduced the Theory of Communicative Action with the main assumption that rationality are of two types: instrumental and communicative. Instrumental rationality is oriented towards success while communicative rationality is oriented towards understanding. The mediatory role of communicative action between the rationalized system and the isolated self is the fact that while the social system aims to efficiently produce outcome (instrumental rationality) the individual attains understanding (communicative rationality) of himself or herself and the system if communicative actions produce its due outcome. And communicative action finds rationality if an agreement is achieved. Thus the end of communicative rationality is consensus and that is the objective of communicative action. Communicative action, in order to produce consensus, should contain the elements of truth, rightness, sincerity and comprehensibility (Roderick 1986: 110-111).

What Habermas is saying is this: freedom... that is freedom instrumentally cannot be completely achieved but what can be attained completely is freedom communicatively. Once an individual enters into a social system for he or she is condemned to belong because he or she is a social being, then he or she should really divest his or her greatest investment in this social system and that is his or her freedom in order to make it work efficiently. He or she cannot just join a group and do whatever he or she pleases. He or she should eventually conform or constrain himself or herself for the system to attain its success... and that is instrumental success. He or she could feel himself or herself liberated if he or she finds consensus with all the instrumentalities and consequences of the social system. Achieving consensus is a way to find the self located and not isolated. Thus an individual would have found freedom... and that is communicative freedom.

But this view is exactly what the first generation of critical theorists objected. Communicative rationality bringing about freedom is illusory. It does not liberate the self. It becomes only a deception. On this note, another epoch comes to life which in itself has a direct focus on the liberation of the body. This is post-modernity which will be tackled after an interlude with logical positivism.

The Birth and Death of Logical Positivism

The fear of reverting back to the dejected state of science was the response of the Vienna Circle in 1922 towards propping up Logical Positivism as a method of assessing scientific practice. This was the other road that led Post-enlightenment to its mourning. The Vienna Circle was a movement designed to eradicate metaphysics as a method of obtaining knowledge and instead intended to pursue knowledge through the scientific method. But with the pursuit of the scientific method came with it the cleaning up of all other methods in order to measure up with what is truly scientific. That is, removing what is not scientific which claims to produce knowledge. The movement was faithful towards keeping up with the scientific practice achieved during the enlightenment period and doing more by proposing methods and rules to find out if theories as the product scientific practice stand out to the truth they seek to discover.

The circle was a loose organization headed by Moritz Schlick (physicist) and composed of luminaries in different fields, Rudolf Carnap and Herbert Fiegl and Fredrich Waismann (philosophers), Kurt Gödel (mathematician), Victor Kraft (historian), Felix Kaufmann (lawyer) and Otto Neurath (sociologist, economist). But this organization got enmeshed and fell victim of the excesses of the Nazi period especially when its leader, Schlick was murdered in 1936. The members of the movement migrated to either United Kingdom or United States, transforming the international movement into Logical Empiricism (Redman 1993:7).

But whether logical positivism or empiricism, they both have the same claim. The primary objective of logical positivists was to act as a vanguard against unscientific enterprise. Since science according to their view is deductive, then the job of a philosopher of science is to examine theories if they could measure up to the truth that the scientific enterprise would like to arrive at. Does this theory "really" produce the results it seeks to explain? The attack was on the language. The initial salvo was on the expression since theory is language and expression.

Science cannot be condemned to the province of daydreamers. Science has a method of achieving knowledge and has logic as a means of expression and validation. Remember, Aristotle began the idea that "truth has a structure." The form of expressing truth is by syllogism. But metaphysics as the means to advance knowledge has to be killed. This would mean exterminating also the theoretical expressions of scientists which do not conform with experimental or observational confirmation, for which, logical positivists claim to have metaphysical traces. Science is armed with either experiment or observation as the method. If the language is theory, then the theory which is the expression of the scientist has to conform with experiment or observation. Anything that a scientist claims in his theory beyond experimental or observational confirmation is metaphysics according to logical positivists.

> Throughout the latter half of the nineteenth century, embryology was at the forefront of experimental research. Among the most important of embryological experimentalists was Hans

Driesch. Two striking laboratory discoveries are associated with his name. Working with sea-urchin eggs and embryos, he was able to demonstrate that the physical deformation of the egg and the subsequent rearrangement of the blastomeres – the cells produced in the first few stages of fission – had no effect on the normal development of the embryo. This experiment suggests that spatial relations among early blastomeres are irrelevant to normal development. Even more strikingly, Driesch went on to show that a single blastomere isolated from the rest at the two – or four – cell stage can give rise to a complete sea urchin embryo normal in every respect except size (Rosenberg 1999:11).

Driesch empirical findings were remarkable but his explanation of the embryos' capability to develop into more complex units from simple ones was marked with incredulity. He proposed that there is an organizing principle he dubbed as "entelechy" a nonmaterial, nonspatiotemporal mode of existence at work in the development of the embryo (Rosenberg 1999:12). It seems Driesch overdid his claim. The nonmaterial, nonspatiotemporal mode of existence cannot be empirically proven since it is non-observable, therefore it cannot be claimed. This is what logical positivists would brand as a form of metaphysical claim and should be eradicated.

In order to accomplish this, logical positivists developed a system to define and validate theories. They advanced the Received View of Theory, which in its original form:

> A scientific theory is to be axiomated in mathematical logic (first-order predicate calculus with equality). The terms of the logical axiomatization are to be divided into three sorts: (1) logical and mathematical terms; (2) theoretical

terms; and (3) observation terms which are given a phenomenal or observational interpretation. The axioms of the theory are formulations of scientific laws, and specify relationships holding between the theoretical terms. Theoretical terms are merely abbreviations for phenomenal descriptions (that is, descriptions which involve only observational terms) (Suppe 1999:18).

A scientific theory, therefore, is composed of mathematical statements and theoretical statements that find congruence with observation. For example, Newton's First Law of Motion states $F = ma$ as the mathematical statement and theoretical statements that find compliment with experience is that a body moves when applied with external force and will continue to move as the force is sustained but will stop in its motion once another force is applied. The mathematical statement is irrefutable but the theoretical terms need empirical proof. In order to solve this, logical positivists devised correspondence rules for theoretical terms to fitly screw in the theoretical statements with observational terms.

$$Tx \equiv Ox$$

The logical positivists strictures would like to get rid of unscientific claims disguising themselves as scientific. The only means for which theoretical statements will pass the gauntlet $Tx \equiv Ox$ is if the empirical evidences can find confirmation in experiments or observation. But there are some serious problems to these. Charles Darwin's Theory of Natural Selection which assumes that "through adaptation, the most fitted in the environment will survive to perpetuate the species and present themselves to evolutionary changes" was accepted as a scientific theory but was devoid of mathematical statement - only theoretical ones. On the other hand, when quantum physics made its breakthrough, there were only accepted mathematical equations to the theory but nowhere can observation verify the very subject of the theory since an atom, much less, the electron can never be seen. If logical positivists will adhere to the strictures of their correspondence rules and verification method, then even the theories that they accept as scientific will not pass the test. Direct verification then, the very criterion of empirical testability of theoretical validation had to give way to a softer method which logical positivists call indirect confirmability.

> A statement is scientifically meaningful if an only if there is actual or possible empirical evidence that tends to confirm, though perhaps not completely verify, the statement (Rosenberg 1999:13).

The phrase "possible empirical evidence" for confirmation has greatly weakened the logical positivist's position. Experimentation and observation are actually experiential or sensory actions. One challenge that logical positivists would have to surmount is actually the meaning of the terms that scientists would use in their theoretical construction. In order to purify the theoretical language that they will use, logical positivists adhere to the following dictum since the verifiability criterion was adopted:

> "The meaning is its method of verification" (Suppe 1999: 21).

This means that the terms that scientists should use in their theoretical constructions should already contain observational entities or simply put, the theoretical statements should have observational content. For example, mass which is a theoretical concept and defined as the amount of matter in a body, is observable by the very definition of the term by weighing the body and the quantity is registered in the scale. The meaning of theoretical terms give rise to operationism.

Operationism, in its fundamental tenets, is closely akin to logical empiricism. Both schools of thought have put emphasis on definite experiential meaning or import as a necessary condition of objectively significant discourse, and both have made strong efforts to establish explicit criteria of experiential significance. But logical empiricism has treated experiential import as a characteristic of statements – namely, as their susceptibility to test by experiment or observation – whereas operationism has tended to construe experiential meaning as a characteristic of concepts or of the terms representing them – namely, as their susceptibility to operational definition (Hempel 1999:45).

The operational meaning and operational definition suggest that the criterion for using certain scientific terms is the outcome derived after certain manipulation of the subject matter has been made. This means that the outcome is experienced, sensed or observed. By virtue of the observational, experimental or manipulative content of the terms which logical positivists espouse as experiential testability of theories, the logical positivists came under attack and one of the strongest barrage according to Suppe (2000) as its leading commando is:

> Correspondence rules were heterogeneous confusions of meaning relationships, experimental design, measurement, and causal relationships some of which are not properly parts of theories (Suppe 2000: S103).

The experiential criterion of experiment or observation became its primary loophole that would bridge the terms used as to the correctness of the theory. With the aim of murdering metaphysics and purifying the claim to knowledge using science, Logical Positivism suffered series of assaults especially when modern physics that was its model of scientific inquiry and which it propped for the social sciences to replicate peered into the things that cannot be seen anymore like the sub-atomic structures. The attack was so great that in March 26, 1969 during the Illinois Symposium on the Structure of Scientific Theories, Carl Hempel, one of Logical Positivism's foot soldier and developer told the audience that he was abandoning the philosophy and on that night, Logical Positivism, according to its antagonists, was buried in its grave (Suppe 2000: S102). With the death of Logical Positivism came also the death of the Received View of Theories.

Popper's Logic

Leading the assault team right at the moment of Logical Positivism's inception in the 1930s was Karl Popper who spoke in the meetings of the Vienna Circle but came not to inspire them but to attack them later. For Popper, science is not a verifying enterprise but a falsifying effort. *The Logic of Scientific Discovery* (1968) is a theory about the scientific method or a theory to account for theories. Popper's task was to provide a logical analysis of the method of empirical sciences. His first volley was to attack the inductive method which David Hume posited to be the procedure of scientific pursuits. His criticism was founded on the thought that theories are not arrived at by building upon singular and particular statements. Such method would hardly end up in the confirmation of universal statements for it would need to employ "inductive principles of higher order" (Popper 1968: 29). What scientists do, in fact, according to Popper is deduction, where, from a universal statement which a theory postulates, singular statements with their own predictive content are deduced and tested to determine if they fitly square with experience, comparing these predictions with the results of practical application or experimental results. This makes a scientist decide on the acceptability of the theory. But such a decision may only be temporary for future negative decision or another theory may overthrow or supersede it. So, as long as the theory could weather the severe test, as science progresses, the theory has "proved its mettle or that it has been corroborated" (Popper 1968: 33). What scientists do then is not to confirm theories but, in the course of testing theory, scientists seek to falsify it. Popper says:

> But I shall certainly admit a system as empirical or scientific only if it is capable of being tested by experience. These considerations suggest that not the verifiability but the falsifiability of a system is to be taken as a criterion of demarcation. In other words, I shall not require of a scientific system that it shall be capable of being singled out, once and for all, in a positive sense; but I shall require that its logical form shall be such that it can be singled out, by means of empirical test, in a negative sense (Popper 1968: 41).

The logic of scientific discovery then is a theory that posits falsification at the heart of scientific endeavor. Science then pursues falsification of theories and examines if they could stand up to the test. From here Popper begins by assuming that "science is not a system of concepts but rather a system of statements" (Popper 1968: 35, 59) and that scientific theories are universal statements, where like all linguistic representations they are systems of signs and symbols. From this universal statement, singular or basic statements are deduced. These statements, however, are empirical for they are falsifiable. A statement like the glass of water is cold is empirical for it can be refuted if a condition is found to the contrary. On the other hand, universal statements can also be framed in a negative fashion. A claim that all mammals have mammary glands or all animals with mammary glands are mammals can also be translated into, "no animals without mammary glands are mammals." These scientific statements, empirical that they are, are tied up with the logical rules of falsification.

> A statement X is said to be 'falsifiable in a higher degree' or 'better testable' that a statement Y... $F_{sb}(X) > F_{sb}(Y)$, if and only if the class of potential falsifiers of X includes the class of potential falsifiers of Y as a proper subclass.
>
> If the classes of potential falsifiers of two statements X and Y are identical, then they have the same degree of falsifiability, i.e. $F_{sb}(X) = F_{sb}(Y)$.
>
> If neither of the classes of potential falsifiers of the two statements includes the other as a proper subclass, then the two statements have non-comparable degrees of falsifiability (Popper 1968: 115-116).

The second rule is logically self-explanatory for having the same number of falsifiable statements, the two theories are explaining the same phenomenon, or that they are one and the same theory. Rule no. 3 would presuppose that without falsifiable statements that a theory would hold as regards another, then the two theories may be explaining two different phenomena and therefore are non-comparable. Rule no. 1 however, provides the specific guide on how to judge one theory to have a higher degree of falsifiability, if one theory has more falsifiable statements than the other.

> We can now say of a theory, provided it is falsifiable, that it rules out, or prohibits, not merely one occurrence, but always at least one event. Thus the class of the prohibited basic statements i.e. of the potential falsifiers of the theory, will always contain, if it is not empty, an unlimited

number of basic statements... Let us now imagine that the class of all possible basic statements is represented by a circular area. The area of the circle can be regarded as representing something like the totality of all possible worlds of experience, or of all possible empirical worlds. Let us imagine further, that each event is represented by radii... or by a very narrow sector along the radii and that any two occurrences involving the same co-ordinates (or individuals) are located at the same distance from the center. Then we can illustrate the postulate of falsifiability by the requirement that for every empirical theory there must be at least one radius (or very narrow sector) in our diagram which the theory forbids (Popper 1968: 90).

With this postulate, it would be said that the issue of fewer or more potential falsifiers would be a criterion to judge if a theory is better than another if it has more empirical content than the other. The more potential falsifiers a theory contains, the larger the class of basic statements it would refute, the more conditions it would prohibit, the larger its empirical content, and the more it says about the world for it could be better refuted by experience. This would mean that the theory possesses a higher degree of falsifiability (Popper 1968: 113). This statement goes back to Rule no. 1 of falsification. Through this pursuit, science grows and, for that matter, knowledge progresses, for in the process of falsification, a theory which has 'proven its mettle' may have been corroborated. Thus Popper dumps verification for a theory is not verified but corroborated. Neither would science be in search for truth for one theory could overthrow another.

> Science is not a system... which steadily advances towards a state of finality... It can never claim to have attained truth (Popper 1968: 278).

Science and the growth of knowledge progress through constant test. New experiences would crop up that would either create a hypothesis fit to falsify a theory or another theory containing more falsifiers could emerge that could rival that theory. With this, one theory could be edged out of contention. Scientific inquiry, therefore, does not end.

> Science does not rest upon solid bedrock. The bold structure of theories rises... above a swamp. It is like a building erected on piles. The piles are driven down from above into the swamp, but not down to any natural or 'given' base; and if we stop driving the piles deeper, it is not because we have reached firm ground. We simply stop when we are satisfied that the piles are firm enough to carry the structure, at least for the time being (Popper 1968: 111).

Popper's objection against logical positivists is the criterion of testability. Popper argued that it is not experiential confirmability or verifiability through experiment or observation that would make one theory pass the test. It is the theory's survival against its confrontation with another theory.

$$T_1 \to T_2$$

Scientific inquiry then is not to meant verify theories but to find out if it would fail. You don't prove great ideas. You try to disprove them.

Kuhn's Revolt

For Thomas Kuhn, science takes a different turn. Kuhn recognized a pattern that underlies the emergence of scientific discoveries. In his *Structure of Scientific Revolutions* (1970), he figured out that the development of scientific enterprise is structured by a period of normalcy; crisis, revolution and back to normal science. Science, according to Kuhn, is characterized with the "uncritical attitude of scientists toward the most fundamental theories and concepts accepted in their field" (Rouse 1987: 27). For Kuhn, these concepts and theories are embodied in a paradigm:

> This remark should begin to clarify what I take a paradigm to be. It is in the first place, a fundamental scientific achievement and one which includes both theory and some exemplifying applications to the results of experiment and observation. More important, it is an open-ended achievement, one which leaves all sorts of research still to be done. And finally, it is an accepted achievement in the sense it is received by a group whose members no longer try to rival it or to create alternatives for it (Kuhn 1972: 88).

Paradigms are sets of theoretical doctrines that constitute the research tradition of certain practitioners, guiding them on what to study, how to go about studying them and what results they would expect. Normal science is paradigm-based research. Experiment and research achieve the results predicted by the theory. Paradigms, however, are closely associated with the scientific community that uses them. But paradigms are limited to the regularities which the theory tries to give account of. Anomalies may crop up that would produce puzzles as to how they could be addressed. These puzzles produce crises in a scientific community that does not only besiege the practitioners but more importantly the theory itself.

> All crises begin with the blurring of a paradigm and the consequent loosening of the rules for normal research. And all crises close in one of three ways... Sometimes normal science

ultimately proves able to handle the crisis-provoking problem despite the despair of those who have seen it as an end of an existing paradigm. On other occasions, the problem resists even appreciating radical new approaches. Then scientists may conclude that no solution will be forthcoming in the present state of their field. The problem is labeled and set aside for the future generation with developed tools. Or, finally, the case that will most concern us here, a crisis may end with the emergence of a new candidate for paradigm and with the ensuing battle over its acceptance (Kuhn 1970: 84).

At the height of crisis, alternative approaches may develop to account for anomalies. The disaffection of some practitioners regarding the unaccounted anomaly may lead practitioners to shift to new paradigms that would provide more convincing solutions to those anomalies. This is called paradigm shift.

> The resulting transition to a new paradigm is scientific revolution... Confronted with anomaly or crisis, scientists make a different attitude toward existing paradigms and the nature of their research changes accordingly (Kuhn 1970: 90).

After the scientific revolution, normal science again proceeds. A competing paradigm may provide different methods of looking at the same phenomenon, new calculations could emerge, and new interpretations could surface. New allegiances and conversions could shift and a new dimension in solving the puzzle could crop up. For Kuhn, therefore, progress in science does not proceed in a textbook manner.

> The transition from a paradigm in crisis to a new one from one which a new tradition of normal science can emerge is far from a cumulative

process, one achieved by an articulation or extension of the old paradigm. Rather it is a reconstruction of the field from new fundamentals, a reconstruction that changes some of the field's most elementary theoretical generalizations as well as many of its paradigm methods and applications. During the transition period, there will be a large but never complete overlap between the problems that can be solved by the old and by the new paradigm. But there will be a decisive difference in the modes of solution. When the transition is complete, the profession will have changed its view of the field, its method and its goals (Kuhn 1970: 84-85).

On account of the classical Newton and the relativistic Einstein, a paradigm shift occurred. Kuhn's scientific revolution could illustrate the exodus of some practitioners from Newton's camp to Einstein's garrison where different approaches to research could ensue and two different ways of looking at the same phenomenon could be deduced. Newton looked at the falling apple while Einstein looked at the indention on the apple surrounding the stem. Newton saw gravity as a string that keeps the planet in orbit with the two masses pulling each other while Einstein provided a description of gravitational field like a bed that indents with the presence of mass on it. As Newton sees gravity as attractive force acting at a distance, Einstein sees gravity as the warping of space-time. Newton illustrates gravity on account of two bodies attracting each other, indicating the calculations of their mass. Einstein describes gravity with the bending of light due to curved space-time. Here, then, are two theories which look at the same phenomenon but see it at two different angles... two theories with two different interpretations of nature.

Kuhn vs. Popper

From here, we ask, how can these two accounts of science be compared? A summary of this comparison is in Table 6.1. Both Kuhn's Structure of Scientific Revolution and Popper's Logic of Scientific Discovery have the same path to cross as they give account of how science progresses or how knowledge grows. But how this growth progresses is where they separate. While both accounts recognize the central role of theory in science, Kuhn's scientific revolution lays the structure of scientific development within the framework of normalcy-crisis-revolution until normal science is again achieved. In this view, scientific revolution is a pattern of theory development, where Kuhn used history to account for such progression. History, however, is not examined in terms of its time element but by virtue of its phases or stages where normalcy, crisis and revolution could take place. For Kuhn, therefore, history and philosophy in science are inseparable.

On the other hand, Popper's Logic of Scientific Discovery champions the use of logical rules not to account for the theory's development but to give account of its appraisal. Such appraisal does not utilize inductive rules but deductive processes where scientific theories are subjected to falsification and are assessed if they could stand up to the test. For this, theory is not identified with practitioners. In contrast, Kuhn asserts that theory is identified with practitioners who constitute scientific communities. These communities subscribe to certain paradigms that comprise the subject of study, the theoretical framework of how they would be guided and the methodology of how they would do the investigation. In this respect, anomalies which the theory could not explain bring about crisis in the community. This would create a revolution in time for a new theory to emerge, which would attract its own following. For Popper, on the other hand, anomalies arise when theories suffer setbacks on account of one theory having more falsifying or prohibiting statements than the other. In this regard, theory appraisal is done through the execution of rules of falsifying basic statements logically deduced from the universal statement derived from the theory. We could demonstrate this in the classic confrontation between Einstein's relativity and Newton's mechanics in regard to gravity.

Table 6.1 Comparing Kuhn and Popper

Categories	Karl Popper Logic of Scientific Discovery	Thomas Kuhn Structure of Scientific Revolutions
Concern	Scientific progress Growth of knowledge	Scientific progress Growth of knowledge
Theme	Method of theory appraisal	Pattern of theory development
Role of practitioners	Theory is identified with practitioners but is used, falsified and corroborated by virtue of the practitioners' conviction of how it could stand up to the test.	Theory is identified with practitioners forming scientific communities
Role of crisis	Crisis comes through the emergence of falsifiable statements or hypothesis that the theory could not give account of.	Crisis emerges through the occurrence of anomalies that the theory could not account.
Emergence of another theory	Theory is appraised through logical rules of falsification	Theory surfaces in confrontation with another theory through a revolutionary fashion.
Which theory is better?	One theory may be better than the other based on its empirical content or if it contains more falsifiable statements.	One theory cannot be judged better than the other.
What science is not?	Science is not verification but falsification	Science is not cumulative but revolutionary
What science is?	Science is a method	Science is a practice

Since the General Theory of Relativity could falsify the statement which the Universal Law of Gravitation could also falsify and the General Theory of Relativity having more falsifiable statements or more statements that it could prohibit, then Einstein's theory has more empirical content and can be judged to be a better theory than Newton's (see Table 6.2). For Kuhn, however, an assessment of which theory is better does not apply for in the course of the revolution to normalcy, certain practitioners may only switch allegiance to another theory and create their own following within the community. That would mean that the other theory is still useful in some respect and other practitioners may still use it in their investigation. Kuhn, therefore, says that science does not progress in a textbook or cumulative manner (Kuhn 1970: 2-3). It is marked with revolutions that convict a group of practitioners to shift paradigms due to the theory's problem-solving capability. For Popper, on the other hand, science is a procedure of testing theories and assessing their fit not only to experience but also to the rules of logic on which theoretical systems are built. It is not verification but falsification that scientists are engaged in. Science then for Popper is a method but for Kuhn, it is a practice.

Table 6.2 Comparison between Newton and Einstein

Statements	Newton's Universal Law of Gravitation	Einstein's General Theory of Relativity
Universal statement	I. All bodies in space move around another body in space due to the force they exert on each other.	II. All bodies in space-time move around the other body on account of warped space-time.

The Milieu-of-the-text

Statements	Newton's Universal Law of Gravitation	Einstein's General Theory of Relativity
Falsifiable statements	a. No body with bigger mass could move around the body with lesser mass since the bigger mass has greater force to pull the lesser mass.	a. No body with lesser mass could indent space-time than the body with greater mass to make it move around it. b. No body in space-time could be in absolute rest. c. No body in space-time could travel close to or beyond the speed of light. d. No path of light or electromagnetic wave could travel in a straight line for it would be bent in the neighborhood of massive bodies.

Lakatos's Counter-attack

From here, Imre Lakatos launches a counter-revolution. After Popper took a beating in the advent of Kuhn's attack, Lakatos mounts his own revenge by salvaging what was left of Popper's logical method. His strategy was to attack Kuhn and then deliver his final blow by advancing his own account of scientific growth.

His primary criticism of Kuhn is that Kuhn read only one dimension of Popper, which is objectionable also for Lakatos. But Popper, he claims, can be appreciated in three different ways. The distinction being: Popper$_0$ of his naïve or dogmatic falsification, Popper$_1$ with his methodological falsification and Popper$_2$ in his sophisticated falsification (Lakatos 1968).

Naïve that it is, Popper$_0$ claims that a theory can be falsified through facts or facts can prove a theory false. Lakatos applauds Kuhn in rejecting naïve falsification since Lakatos claims that facts cannot just debunk a theory. Facts also have to be interpreted. A theory is made of harder stuff that not even plain observables could prove it wrong. One good example is the sun that rises in the east, climbs up above at noonday and sets in the west. It is an apparent observable. The sun, by plain observation, revolves around the earth. But in reality that is not so. The observable cannot debunk Newton's Law of Gravitation which says that it is the earth that revolves around the sun.

Popper$_1$, with his methodological falsification, asserts that propositions crafted in a mode fit for methodological deduction can refute a theory. Lakatos also finds this objectionable since he claims, "Propositions can only be derived from other propositions (Ball 1987: 21)." Propositions do not stand alone. They have firm theoretical basis. According to Popper$_1$'s logic of research, "if there is a clash between a low-level, 'observational' hypothesis and a higher-level theory – shown in his 'deductive model' – the theory must be rejected (Lakatos 1968: 154)." In a deductive model, one can proceed through this:

> No object of lesser mass in space can exert force to a body of bigger mass in order to make it move around it.
>
> The moon is a body of lesser mass than the earth.
>
> Therefore, the moon is an object that cannot exert force to the earth which is a bigger body in order to make it move around it.

The Milieu-of-the-text

This is the logical form using Newton's Law of Gravitation and applying it to the earth and the moon. But how do we apply a falsifying proposition? Let's take this example.

> No object moves unless acted upon by an outside force which displacement can be measured by dividing the distance of displacement over the duration of displacement.
>
> Light is an object.
>
> Therefore, light moves unless acted upon by an outside force which displacement can be measured by dividing the distance of displacement over the duration of displacement.

The conclusion then becomes anomalous. It would mean that we can divide any distance with the time of displacement pertaining to light and come up with its speed. We can have 1,000,000 miles as distance and 1 second as the time, making the speed of light's displacement to be 1,000,000 miles per second. This is highly anomalous due to one falsifying statement.

> No body can travel beyond the speed of light at 186,000 miles per second.

This falsifying proposition does not stand alone for it is lifted from the Einstein's Special Theory of Relativity. Even if the logic of the previous syllogism would be anomalous using the falsifying statement, still this does not prove Newton's First Law of Motion to be wrong. It may only be inconsistent to the condition which light presents itself or may prove inapplicable under those conditions.

While disputing $Popper_0$ and $Popper_1$, Lakatos rescues Popper in his $Popper_2$ version of scientific progress. $Popper_2$ provides the criterion to judge which theory is better than the other by adding sophistication to the methodology. Popper$_2$'s dictum states that:

> A theory is better than its rival (a) if it has more empirical content, that is if it forbids more 'observable' states of affairs, and (b) if some of this excess content is corroborated that is, if the theory produces novel facts (Lakatos 1968:163).

The confrontation between Newton and Einstein illustrates how sophisticated falsification proceeds (Table 6.2). Sophisticated falsification does not, therefore, proceed using single theory isolated from the rest. It is always done in comparison with other theories.

> Now let us call a series of successive theories each of which is acceptable - that is, each of which has higher content than its processor – a (theoretically) progressive problem-shift (Lakatos 1968, 164).

Now how does this problem-shift occur? From here, Lakatos proposes his own concept of research program in his *Theory of Scientific Growth* (1968).

> Scientific progress, according to Lakatos, can only be gauged by looking at the successes and failures, not of single theories but of successive series of theories, each sharing common core assumptions. Such a series he calls "research program." A research program consists of a "hard core" of not-directly-critizable assumptions. The hardness of the hard core is assured by the programs negative heuristic, that is, the methodological rule that criticism be directed away from the hard core of the program. The program's positive heuristic, by contrast, prescribes the construction of a protective belt of auxiliary assumptions and hypotheses which serves to protect the program's hard core. It

is this protective belt, Lakatos says, which has to bear the brunt of tests and get adjusted and readjusted or even completely replaced, to defend the core. A research program is successful if all this leads to a progressive problem-shift; unsuccessful if it leads to a degenerating problem-shift (Ball 1987: 24).

A progressive problem-shift is achieved if it leads to the discovery of novel facts or that it contains more empirically corroborated content. The theory degenerates if the problem-shift has not successfully surmounted falsification and no new novel fact nor more empirical content has been generated. Lakatos does not stop at simply giving account of this sophisticated falsification but he also lays down the practice of scientists on how they defend their research programs and hold on to their theories by adjusting the positive heuristic while preserving the untouchable assumptions of the negative heuristic in the advent of scientific experiments.

> Let us take Newton's mechanics and the law of gravitation (the hard core C of the programme); and the initial conditions in some planetary system and several observational theories (the protective belt B_1). Let us imagine that a planet p slightly disobeys the theory N_1, made up from C and B_1. Would the Newtonian consider that this refutes C? NO. He will suggest changing the hypotheses, say, about the initial conditions and will suggest that there must be a hitherto unknown, very small planet, p. perturbing the orbit of p. He propose an auxiliary theory of p' describing its orbit, mass, etc. Then he will test the proposed orbit of p', replacing B_1 and B_2. He would try to plan bigger telescopes to make this conjectural orbit p' discernible, testable. But if it seems that the conjectured planet is not in the reach even of the

biggest optical telescopes, he may try some quite new instrument (like a radiotelescope) in order to enable us to 'observe it', that is k to ask – however indirectly- Nature about it. The new observational theory may itself be poorly articulated, but fir the time being they will care. If the new instrument locates the planet where C and B_2 predicted, the result will be hailed as a victory for the research-programme (arrived at by sacrificing B_1 and incidentally, also for the new observational theory (Ball 1987: 24).

This demonstrates progressive problem-shift and Lakatos's example actually occurred with the discovery of Neptune as it perturbed the orbit of neighboring planet, Uranus. But even with the success of Newton's mechanics, its accolade met serious question with the new problem posed by Special Relativity. The Theory of Special Relativity assumes that the speed of light, which is constant at 186,000 miles per second, is unmatchable. Let's assume this is the hard core C of the theory. It assumes, therefore, that the speed of an object can be based on the speed of light. Since the speed of an object can be gauged through the speed of light, then time can be also be measured in relation to the speed of light and distance can be calculated using the same speed as basis. The unity of space and time has now been attained. Since an object can never move faster than the speed of light (C_1), making the speed finite (C_2) then, with the unity of space and time, two consequential hypotheses could be derived:

Time is not infinite (H_1). And space is also not infinite (H_2).

Newton's mechanics assumes that distance or space is separate from time. Thus you can divide one from the other and arrive at the measure of speed. Consequently, you can then divide any amount of space with any amount of time and arrive at the speed of the object. Special Relativity does not agree with this, for time and space are not infinite since both variables will depend on the motion of the object in relation to the speed of light. Now to prove this, a scientist can make an experiment. Let's say he rides a bus carrying a separate stopwatch while his colleague who is also holding his own stopwatch, remains at the bus station where the bus starts its journey. They synchronize their watches and after the trip they compared the time their stopwatches ticked and still there is not one difference. If time were relative to the motion of the object, then the one who made the trip should have logged in at a different time. Does this mean that Special Relativity is wrong? "No," says Lakatos. It is because the speed of the bus is way below the speed of light.

Another experiment can be embarked on. Let's say, rather than riding on a bus, the scientist rides a rocket and hurled himself in outer space which left his colleague with a stopwatch ticking at the same time at the launching pad. Since the rocket flies at greater speed, a difference in the time is registered upon the rocket's return. The watch brought inside the rocket is a little delayed than the one that was left on the launching pad. With this, H_1 is confirmed. Progressive problem-shift then has been attained, preserving the core assumptions of the theory and confirming the peripheral hypothesis created by the theory.

Einstein pushed this theory even further. He claimed that Newton's Law of Gravitation does not square with the core assumption of special relativity. A simple thought experiment would confirm this. Since Newton's Law of Gravitation assumes that gravity is force equivalent to the multiplication of the mass of the two bodies over the square of their distance, then force would also be infinite as to the distance separating them. If, for example, we have the Sun and the Earth as two bodies in space, then we suddenly pull the Earth away from the Sun, the loss of gravitational force would be instantaneous. This cannot be, according to Special

Relativity since it would appear that the speed by which the effect of the force of the Sun acting on the Earth would have vanished faster than the speed of light.

With this, Einstein generalized the theory of relativity to include gravity which assumes that gravity is the warping of space-time in the neighborhood of mass or energy. This then becomes the core assumption C of the theory. Since it assumes that space-time is bent near a massive body, then light will also be bent near it. That being the case, then an experiment can be done. You can suspend a ball of a certain mass in the air and observe a ray of light being shot near it while using two mirrors to deflect it in a triangular manner around the ball and back. Normally the sum of all angles of the triangle would measure 180^0 if there were no curvature around it. And really, with this set-up no noticeable difference from 180^0 sum would be registered. Does this mean that the theory is false? Lakatos says "no," for the ball has a mass so small to bend it. The experiment would be noticeable to larger bodies in space. The better experiment would be observed as regards the Sun being massive enough to bend the light of a star near it. But it can only be confirmed in solar eclipse since the brightness of the sun could hide the brilliance of the star. This was actually confirmed in 1919.

> Taking advantage of this fact, the test was actually made in 1919 by a British astronomical expedition to the Principe Islands (West Africa), from which the total solar eclipse of that year could best be observed. The differences of angular distances between the two stars with and without the sun between them was found to be 1.61" ±0.30" as compared with 1.75 predicted by Einstein's theory (Gamow 1961: 110).

Again, this has proven to be a progressive problem-shift as exemplified by the victory achieved by the theory to have survived the test.

Popper vs. Kuhn vs. Lakatos

What led to this debate is the epistemological thirst to blueprint the growth of knowledge where the scientific enterprise occupies a unique position. And what is amazing in this enterprise is its capability to reproduce what is to be known. This known thing emerges out of a method that is able to capture, explain and even predict. But there are a few things that move right into the heart of how the theory is able to bring about the known thing. Given rival theories that are able to bring about the known things, which one is better? How can we give account of the emergence of these amazing statements that are able to capture and uncover the supposed thing to be known?

The desire to give account of how theories can capture what is to be known and how one is better than the other in the advent of competing theories is in itself theoretical. The account to explain theories as the enduring ingredient of science also brings about theories to account for them. In explaining the development of theories, we can also give account of the progress of science, thereby satisfying our thirst to account for the growth of knowledge.

For Popper, science does not proceed by accretion. Theories, which lie at the heart of science, are not advanced after fitting in and filing over of proven hypothesis. Rather, it proceeds out of bold and creative assertions. These assertions, inventively crafted in a system of statements, are then shot at with hypotheses that would attempt to prove them false. Thus, instead of verification, the scientific enterprise is a process of falsification. Since theories are proposed, scientists embark on experiments which are intended not to prove them right but to find them false. Science is, therefore, to dispute or refute. If experiments were done to prove or verify them, then the practice of science is meaningless for what is actually achieved if a theory is simply given for verification. A scientist who conceived of the theory is convinced it is correct. Every scientist or theorist would always presume his theory works. The next thing is to shoot it down. But if falsification fails, then the theory has proven its worthiness.

The process would not be inductive but deductive as governed by logical rules. But the logical process is not that simple for one hypothesis, although it is proven not to square with the theory would not outrightly prove the theory wrong and make the theory a candidate for extinction. The brilliant theorist could adjust or fine tune some problematic assumptions of the theory and render it still workable.

In his Theory of Scientific Discovery, the method of falsification and logical deduction could provide the bases of which theory could be adjudged better than the other and if a series of falsifying statements could really refute the theory without any hope of fine tuning it, then theory has met its last gasp of air.

But even with the method of determining better theories as against its rivals, the scientific enterprise would appear discontinuous for the theory can meet its death (Table 6.3). On the other hand, since Popper assumes that science proceeds out of bold assertions, then theories are proposed out of bold assertions that just prop up without one building from the other. After one theory is proposed, science would then await for another brilliant assertion which can give account of the thing that is waiting to be known.

Kuhn, however, saw a pattern on account of these bold assertions. These brilliant assertions, he says, are themselves revolutionary that could shake the practitioners of science and could rival well-entrenched theories. The pattern, he says, is a revolution which these bold assertions could create and after which the scientific community experiences normalcy upon its acceptance by another group of practitioners. Science then awaits another bold assertion to shatter the normalcy and spur another revolution. For Kuhn, "which theory is better" is not the issue. Science is not about falsification. A fact or a falsifying hypothesis could not overthrow a theory since a fact is still subject to interpretation and a hypothesis simply emanates from another theory which could itself be falsified if the method is used.

Table 6.3 Comparing Popper, Kuhn and Lakatos

Categories	Popper	Kuhn	Lakatos
Main Concern	Falsification	Scientific Revolution	Scientific Growth
Theory	The Logic of Scientific Discovery	The Structure of Scientific Revolution	The Theory of Scientific Growth
Unique Claim	Logical deduction or Falsification	Scientific Paradigm	Research Programme
Scientific Enterprise	Science is a method	Science is a practice	Science is both method and practice
Scientific Progress	There is a method to assess scientific progress	No method to assess scientific progress	There is a method to assess scientific progress
Continuity of Science	No continuity	No continuity	There is continuity in the scientific enterprise

In fact, two rival theories could both work and contain definite explanatory value only within the grasp and boundaries of their elemental assumptions. This would then permit the emergence of paradigms which lay down the specific units that they study, the set of limitations that the theory possesses, what the theory could explain and the methodology appropriate for them. It would just be up for the practitioners which paradigm to accept and use. Since science is a revolutionary occurrence, science then with theories as its core element is a discontinuous process. One theory would be shaken with the introduction of another until a period of normalcy among the community of practitioners of science ensues.

Lakatos, however, defends Popper and attacks Kuhn although in the process he was able to marry both Kuhn and Popper. Lakatos subscribes to Popper's falsification and even sophisticates it. In his Theory of Scientific Growth, he agrees with Popper's view that the scientific enterprise is all about falsifying well-entrenched assertions of the theory and not proving them. But falsification is a complex activity involving the brilliant modification of the theory's assumptions that blanket the core claim of the theory. For Lakatos, science is about adjusting and fine-tuning the assumptions of the theory once confronted with falsifying statements. He then proceeds where Popper left off. Popper suggests that theories face the threat of extinction once they crash with a falsifying hypothesis. Theorists would then adjust some assumptions of the theory but Popper did not specify how.

Lakatos, on the other hand, provides the procedure on how this adjustments is done, advancing his concept of research programs. Scientists do not simply abandon the theory once it is confronted with falsifying hypothesis, instead, they adjust the peripheral hypothesis of the theory, making it still workable to account of more empirical evidences. The theory progresses but if the barrage of these falsifying statements are left unchallenged, leaving less empirical evidences to account for, then the theory might meet its death.

Science then is not only a method of falsifying the bold assertions but is in itself, a practice just as Kuhn asserts.

The research program works like a research paradigm that evolves into a more resilient theory that could account for more empirical evidences and give a better picture of the world. Science then is continuous and the progress of the theoretical undertaking comes with it the growth of science.

The Postmodern Experience

Habermas could be said to have stood midway between the modernist of his time and the postmodernist that came later. While modernity found its way in Europe with the spread of capitalism at the height of the Industrial Age, postmodernity could be reckoned to have surged its way in a fast changing society marked with high degree of technological advancement, with the accumulation of information and with the attainment of progress in the 1970s. Anthony Giddens in his book *The Consequences of Modernity* outlined that late modernity could be sociologically characterized with:

> Separation of time and space and recombination in forms which permit the precise time-speed "zoning" of social life, disembedding of social systems (a phenomenon which connects closely with the factors involved in time-space separation), and the reflexive ordering and reordering of social relations in the light of continual inputs of knowledge affecting the actions of individuals and groups (Giddens 1990: 16-17).

At no particular time in history can the person in his or her own present converse and create relations with someone in his or her past or future at the same present. This can technologically be done with someone on the phone located in one time zone but talking with someone situated in another part of the earth with either delayed or advanced time. With live video streaming a person can participate in a located activity as it happens even if he or she is in a time zone that is either in the past or future of the person at the other end. The present, past and future can be instantiated at the same present which is made possible with disembedding mechanisms that enjoy the same presentness.

With the advent of modernity, reflexivity takes on a different character. It is introduced into the very basis of a system reproduction; such thought and action are constantly refracted back upon one another... The reflexivity of modern social life consist in the fact that social practices are constantly examined and reformed in the light of incoming information about the very practices, thus constitutively altering their character (Giddens 1990: 38). This is late modernity as Giddens outlined.

Postmodernity, however, appeared in the language of sociologists and philosophers with the publication of Jean-Francois Lyotard's book *The Postmodern Condition: A Report on Knowledge* in 1979. The very first statement of the book strikingly presents its aim as:

> The object of this study is the condition of knowledge in highly developed societies (Lyotard 1979: xxiii).

The phrase the "condition of knowledge in a highly developed societies" attracts its own unique worth for at no time in history has knowledge translated into information become an everyday essential object accessible at no limited space and time. If the period of Enlightenment was a revolution in knowledge, a revolt against the church to dispel myth and ascend science as knowledge, postmodernity has taken scientific knowledge as everyday information that could be gobbled up like daily meal available as one opens the pantry.

The Milieu-of-the-text

> In the postindustrial and postmodern age, science will maintain and no doubt strengthen its preeminence in the arsenal of productive capacities of nation-states... Knowledge in the form of an informational commodity indispensable to productive power is already, and will continue to be, a major – perhaps the major – stake in the worldwide competition for power. It is conceivable that the nation-states will one day fight for control of information, just as they battled in the past for control over territory, and afterwards for control of access to and exploitation of raw materials and cheap labor (Lyotard 1979: 5).

Lyotard, however, explains that knowledge comes in two types.

> Scientific knowledge does not represent the totality of knowledge; it has always existed in addition to, and in competition and in conflict with, another kind of knowledge, which I will call narrative (Lyotard 1979: 7).

With the intersection of these two types of knowledge comes the problem of legitimation or the issue of knowing how that what we know is correct. Legitimation is a problem of who to judge and how to judge that a claim is correct in a society where the social bond is created out of communication. Lyotard calls this language games. A language game consists of a sender (the person who issues the statement), its addressee (the person who receives it) and its referent (what the statement deals with) (Lyotard 1979: 9).

Lyotard presents a narrative like that of a myth or legend in a primitive society where the old people are the sender of source of the narrative, the young people are the addressee and the referent being the content of the narrative. The narrative is legitimized or presumed to be true not mainly for its truth but for its moral-value through their social bond. The legitimating process is subscribed with the factors of respect, honor and authority ascribed to the old people who are the only legitimate authorities in transmitting the narrative. The young generation possesses no moral authority to transmit the narrative though they can narrate it but the addressee or other young people would also ask who is their source and in turn refer to the senior members of the society in order to achieve some approbation. The specific roles that the sender and addressee with respect to the referent possess are constant and unchanging. The addressee cannot be the sender and still possess the legitimacy over the referent. The narrative is legitimized through their social bond and the narrative solidifies the bond. Here, Lyotard fortifies this concept of language game for like a game, this solidifying form of communication has rules inherent in the game. Just like any game, no one can play the game unless he or she accepts the rules. For the primitive society the narrative is in itself knowledge that travels only one way from the sender to the addressee. That is the inherent rule.

But the postmodern society runs differently. Since scientific knowledge has become informational, the addressee which is the receiver of the scientific knowledge can also become the sender at one given time and his or her addressee can also act as the sender at another given moment as he or she transmits the referent. The scientist who is supposed to be the sender of the referent will end up having many senders as the scientific knowledge is passed on as if the other senders are also expert sources. The problem here is that, the scientist who is supposed to be the source of the knowledge does not stand as an unquestioned expert for he or she could end up being critiqued not only by his fellow scientists but also by the public which at one time has also become senders of the referent. Scientific knowledge then does not travel one way. "Scientific knowledge is a kind of discourse," Lyotard

(1979:3) assumes where the public becomes active participants and not only passive receivers. The scientist-expert does not possess the hegemony of knowledge but he or she will be questioned and criticized by the public which is a collectivity of those who know much, those who know less, and even those who don't know anything. Here is where the problem of legitimation comes in. The Enlightenment period and the age of modernity paddled through their courses with the scientist-expert monopolizing the "talk" and he or she is the only legitimizer of the subject he or she is an expert of. But the postmodern society emerged with everyone else participating in the "talk" that the problem of the legitimizer also becomes an issue. Who is correct and how we know that what we know is correct if everyone else throws questions about it and everyone else answers back? We have come to the point that everyone else questions and talks that the only way to determine what is true is first to doubt. To doubt comes with it the method of breaking apart what we might provisionally believe to be true and breaking it apart technically means to deconstruct.

If the Enlightenment period spells its abhorrence on myth and props up science as knowledge, and while modernity emerged to elevate the question of how the self can be freed in a period where science has created an organized system, postmodernity has run its course to question not only the system but also science itself. This has come with the empowering and increasing autonomeity of the self. Postmodernity assumes that the self can only be freed if it has the power to autonomously determine what is it that he or she is supposed to know. This is realized with the main assumption of another philosopher named Michel Foucault that knowledge is power and power is knowledge.

> Perhaps, too, we should abandon a whole tradition that allows us to imagine that knowledge can exist only where the power relations are suspended and

that knowledge can develop only outside its injunctions, its demands and its interests. Perhaps we should abandon the belief that power makes mad and that, by the same token, the renunciation of power is one of the conditions of knowledge. We should admit rather that power produces knowledge (and not simply by encouraging it because it serves power or by applying it because it is useful); that power and knowledge directly imply one another; that there is no power relation without the correlative constitution of a field of knowledge, not any knowledge that does not presuppose and constitute at the same time power relations. These 'power-knowledge relations' are to be analyzed, therefore, not on the basis of a subject of knowledge who is or is not free in relation to the power system, but, on the contrary, the subject who knows, the objects to be known and the modalities of knowledge must be regarded as so many effects of these fundamental implications of power-knowledge and their historical transformations. In short, it is not the activity of the subject of knowledge that produces a corpus of knowledge, useful or resistant to power, but power-knowledge, the processes and struggles that traverse it and of which it is made up that determines the forms and possible domains of knowledge (Foucault 1975:27-28).

What we know, therefore, is constructed by power for power creates knowledge and power is knowledge and this power has emerged with the use of apparatuses and modalities. For example, order in society has been kept with the institution of the prison and the knowledge of disciplining us is put into use with the strategies of torture, the carcel or the panoptical prison or the concept of panopticism... "the idea of surveillance or that we are being watched has created a disciplined, well-organized society" (Foucault 1975). The method then of emancipating one's self is to deconstruct the modalities of power. Deconstruction was first devised by Jacques Derrida as a critique of Western philosophy with the publication of his book *Of Grammatology* in 1967.

> Deconstruction is generally presented via an analysis of specific texts. It seeks to expose, and then to subvert, the various binary oppositions that undergird our dominant ways of thinking—presence/absence, speech/writing, and so forth. Deconstruction has at least two aspects: literary and philosophical. The literary aspect concerns the textual interpretation, where invention is essential to finding hidden alternative meanings in the text. The philosophical aspect concerns the main target of deconstruction: the "metaphysics of presence," or simply metaphysics. (Reynolds 2016).

Deconstruction is intended to unveil the cemented ways of thinking in order to reverse how such is viewed. Derrida's version of deconstruction is basically textual where he presumed some hidden or unintended meaning behind the text. Foucault's version of deconstruction searches for modalities of power in order to unmask prevailing institutions or ways of thinking. Foucault's deconstruction is a method that historically searches for the apparatuses or strategies of power in order to peel off the layers of what we know or claim to know. The postmodern society then has not just created an autonomously knowing but also an emancipated self who will not just question the inner layer of what he or she knows but also questions the very system which he or she is in. While knowledge has increased in a very fast rate, while access to this knowledge has come at ease, these highly industrialized societies have also come to the phase where they have a difficulty grappling on the problem who to believe and what to believe anymore. Postmodernity has produced a questioning and doubting society where science is not spared from its incredulity.

Postmodernity's Attack on Science

The whole crazy thing about this is that science that built the society is now being attacked by the very society it has built. Reason which radically demolished the church-dominated society of the middle ages and re-ordered it into a science-centered one seemed to have turned against itself. Knowledge that reason produced is now being attacked by reason itself. Science that reason has created to discover the natural world and re-order the society is now being besieged by reason itself. Knowledge that is supposed to liberate turns out to enslave which power-relations have constructed but which reason had to deconstruct again. Science that is supposed to be the fundamental arbiter of what knowledge is, has become too monopolistic that postmodernity objects on its claim that it is the only source and arbiter of what is to be regarded as true. It is here that postmodernity abhors the credulity of metanarratives.

> Postmodernity is defined as incredulity toward metanarratives (Lyotard 1979: xxiv).

The irony of it is that both science and philosophy experiences intellectual orgasm over metanarratives. Science is intellectually ecstatic towards the devising of grand theories. Philosophy is euphorically addicted towards the crafting of universals about the metaphysics, epistemics and ethics of things or social relations. Grand theories and universals are, no other than, metanarratives that postmodernity despise.

Physics is geared towards uniting the four fundamental forces in nature: gravitational force, electromagnetic force, weak nuclear force and strong nuclear force into one theory of everything. It seeks to unite the General Theory of Relativity with Quantum Mechanics. For philosophy, logic would not proceed without crafting universal statements. Metaphysics seeks to universalize the nature of things and epistemology wants to find a general view of knowledge.

> Postmodern critique of scientism goes well beyond mere rejection of its epistemological foundation and metanarrative. Postmodernism is incredulous to all metanarrative because it rejects any form of epistemological foundationalism (Leffel 2000: 51).

Postmodernists argue that there is not just one theory that could monopolize the explanation for everything since the diversity and uniqueness of each one characterizes the "every" of "things." Postmodernity claims that there could not just be one theory to account for everything since everything else is particularly multi-dimensional. But science is a natural hegemon.

Since the publication of Charles Darwin's *Origin of the Species* in November 24, 1859, where the Theory of Natural Selection became the foundation of evolutionary biology, evolution has become a popular idea that occupied most intellectual spaces of most disciplines. Bones and other fossils once exhumed and analyzed by archeologists and paleontologists are viewed as the product of evolutionary forces traceable from the beginnings of time. In 1962, Robert B. Fox discovered a skull cap in Tabon Cave Palawan which was carbon-14 dated to be as old as 30,500 BP (Before Present) (Jocano, 1975: 64).

> It is clear that men were present in Island Southeast Asia as early as one and a half to two million years... At present two choices are open to students of prehistory in Island Southeast Asia, particularly in the Philippines. The first one is to view the peopling of the region as a result of a continuous process of human evolution which started, as evidences show, from Homo erectus

and developed through time, into a dominant core-population of Homo sapiens (Jocano, 1975: 66-67).

This is not to cast any aspersion from evolutionary sciences but evolution has become hegemonic as postmodernists would view it. Diet and the human body's response to it are also explained in terms of evolutionary processes.

> Meat has played a starring role in the evolution of the human diet. Raymond Dart, who in 1924 discovered the first fossil of a human ancestor in Africa, popularized the image of our early ancestors hunting meat to survive on the African savannah. Writing in the 1950s, he described those humans as "carnivorous creatures, that seized living quarries by violence, battered them to death ... slaking their ravenous thirst with the hot blood of victims and greedily devouring livid writhing flesh"... Eating meat is thought by some scientists to have been crucial to the evolution of our ancestors' larger brains about two million years ago. By starting to eat calorie-dense meat and marrow instead of the low-quality plant diet of apes, our direct ancestor, *Homo erectus,* took in enough extra energy at each meal to help fuel a bigger brain. Digesting a higher quality diet and less bulky plant fiber would have allowed these humans to have much smaller guts. The energy freed up as a result of smaller guts could be used by the greedy brain, according to Leslie Aiello, who first proposed the idea with paleoanthropologist Peter Wheeler. The brain requires 20 percent of a human's energy when resting; by comparison, an ape's brain requires only 8 percent. This means that from the time of *H. erectus,* the human body has depended on a diet

of energy-dense food—especially meat (Gibbons 2013).

Recent researches on diet and human evolution has also taken genetics as the primary entry point. Scientists in 2015 sequenced the genome of a Neanderthal woman and a Denisovan girl in a cave in Siberia and found interesting results.

> A key area of interest for the scientists is the gene for taste receptors, which are molecules on taste buds that help people taste flavors. They found that the genes for two bitter taste receptors, TAS2R62 and TAS2R64, mutated in hominins after the ancestors of chimpanzees and hominins diverged, making the hominin versions inoperative. They found that this mutation occurred before the split between the ancestors of modern humans — Neanderthals and Denisovans. It remains uncertain what specific bitter molecules these receptors target, but they may be substances that are common in the diets of most or all great apes, but that are rare or absent from hominin diets..."Since we know these mutations are specific to the human lineage, perhaps we can learn something about human evolution by figuring out what substances the functional versions of these receptors are responsible for tasting," said lead study author George Perry, an anthropological geneticist at Pennsylvania State University in University Park (Choi 2015).

The heaven is not spared. With Einstein's General Theory of Relativity, the universe is understood to be expanding. Reversing the process would result in a universe ending in a single point. This is then the beginning and by beginning it would presume a creator. Stephen Hawking in

1983 debunked the concept of the universe's beginning with his "No Boundary Proposal."

> "The idea that space and time may form a closed surface without boundary also has profound implications for the role of God in the affairs of the universe. With the success of scientific theories in describing events, most people have come to believe that God allows the universe to evolve according to a set of laws and does not intervene in the universe to break these laws. However, the laws do not tell us what the universe should have looked like when it started – it would still be up to God to wind up the clockwork and choose how to start it off. So long as the universe had a beginning, we could suppose it has a creator. But if the universe is really completely self-contained, having no boundary or edge, it would have neither beginning nor end: it would simply be. What place,, then, for a creator? (Hawking 1988: 149).

The No Boundary Proposal subscribes to the assumptions of Quantum Mechanics that at the level of sub-atomic particles, the electron does not appear to be a point but a smear since particle could behave both as particle and wave. So at singularity there is no point but a wave. At this stage, planets, stars and galaxies and even life is presumed to have evolved.

> Some 15 billion years ago the universe emerged from a hot, dense sea of matter and energy. As the cosmos expanded and cooled, it spawned galaxies, stars, planets and life. (Peebles, et. al.1994).

Even beauty is appreciated to have taken its due evolutionary course as preferences for a particular nature of aesthetics is viewed in terms of reproductive fitness. This is couched in the Theory of Natural Selection.

> According to the theory of evolutionary biology, beauty standards are those characteristics that will attract a mate who will pass genes on to the next generation, claims psychology professor Glenn Wilson in the following viewpoint. For example, facial qualities such as symmetry and sexual dimorphism—traits that are clearly male or female—signal reproductive fitness, he asserts. Evolutionary biology also determines standards of body beauty, Wilson argues, as men prefer women with a much narrower waist than hips, an indicator of high levels of female hormones. However, he suggests, culture does account for some variations in beauty standards that relate to social aspects of mate quality, such as wealth. Wilson is a pioneer in evolutionary theories of sex differences, attraction, and love (Gale Document 2013).

While Enlightenment has gotten rid of God to be the answer to everything we know, evolution followed through in the modern and postmodern societies with its own revenge. While evolution is presumed to be scientific, postmodernity loathes it as too monopolistic of its presumption to answer for every beginning.

This brings us to the problem of objectivity. Postmodern thinkers believe that the scientist is not free from ideological biases.

> Postmodernist critique centers on the relationship between social context and scientific theorizing. They hold that no one can access reality from a culturally neutral context. It is not possible to remove scientist from their social and ideological biases. Consequently, researchers interpret data selectively, based on their own social location. Theories turn out to be an extension of the socially constructed consciousness of the researcher and the research community. Consequently, all theory

is socially grounded interpretation rather than transculturally objective truths – foundationalism is at best a pretense, really little more than covert political posturing (Leffel 2000: 52).

Scientists, on the other hand, argue that the use of theory and rigid methodology detaches the scientist from the object of his or her study. But postmodernists rebut this on the note that scientific thinking and scientific method have become a cultural norm in the modern society, that the scientific method has become the only arbiter of what can be accepted as knowledge.

Table 6.4 Postmodernity's Offensives on Science

Science	Postmodernity's Attack
One of the goals is to the unification of theories into a grand theory.	Despises grand theories and universals as it is skeptical of metanarratives.
Achieves objectivity using theory and rigid methodology.	Scientists are still part of the social constructions and are not free from ideological bias.
Language is a tool to convey scientific findings.	Language is a tool of persuasion and power.

If the scientist is not free from ideological biases, the language that he or she uses is not at all neutral. Science argues that language is simply a tool to express the relationship of variables. Postmodernity, on the other hand, retorts that language is power-laden. Scientific pursuits do not proceed from a vacuum. A theory has been crafted in order to confront another theory for the other cannot explain some aspects of the phenomenon. Einstein's General Theory of Relativity addresses the issue of light's constant speed with the warping of space-time which Newton's Universal Law of Gravitation was short of explaining since it views gravity as force of a bigger body acting on a smaller one at a distance.

Newton's Universal Law of Gravitation, on the other hand, refines Galileo's proof of a force acting on a falling body with constant acceleration and proves of Copernicus' geocentric planetary system with his own telescope. Copernicus, however, debunks Ptolemy and Aristotle's sun-centered universe. In all these cases, a theory presents itself as a refinement or total abrogation of another. And language is used to confront the other. In this confrontation, power dynamics emerges.

> Postmodernists are interested in the rhetorical power of language as it relates to scientific theorizing. Words are inherently tools of persuasion and power. For this reason, there is always occasion for a subversive reading of science to reveal a political context. Once the subtext is revealed, the marginalization, the marginalization of the scientific revealed, the marginalization of the scientific metanarrative can be exposed... Linguistic theory is crucial to the postmodern rejection of foundationalism and metanarrative. According to postmodernists, human though is mediated through language – we think and communicate linguistically.... no one can conceptualize reality outside his or her own language system ... The rules of language, syntax and semantics, establish the rules of rational thought (Leffel 2000: 53, 54).

The Milieu-of-the-text

Science cannot proceed without textualization. As defined before, textualization is a bounded enterprise of rigor where the reality that the enterprise seeks, the method to seek for the reality and the rules that govern both the seeking and the subject of what is being sought are all proposed, validated and challenged in constructs that encapsulate the phenomenon as a subject that is meant to be read and comprehended in a process that deconstructs, reconstructs and foreconstructs.

Textualization does not simply mean putting concepts on paper. The milieu-of-the-text goes far beyond "paperizing" concepts. The milieu-of-the-text which textualization has brought about is characterized with certain nature. *First is its ability to capture and reduce.* Since textualization is a bounded enterprise, the milieu-of-the-text is discriminatingly specific to a certain phenomenon. Capturing the phenomenon necessitates describing the extent or boundaries of the phenomenon or setting it apart from others. Reducing the phenomenon is the ascription or invention of terms that would describe the phenomenon. But once captured and reduced, the milieu-of-the-text brings about one unique quality. *The text itself creates the compression of time and space.* It is not singularity which physics refers to time nor the unity of time and space, but it is the squeezing or condensing of time and space into minimal use of words. This is very evident in historical materials. In just one paragraph dated a certain time, a certain event is bounded and described in a few phrases or sentences. In another, the event shifts to a different time frame, with an elapsed time of hundreds or thousands of years. You are going through hundreds of years in time in only one or few sentences with only a few seconds of perusing. You are consuming a few seconds of reading events that transpired thousands of years in difference. The same thing happens in space. You may be going to and from space separated by hundreds or thousands of miles in only one or few sentences. This is one quality of the text which is able to compress time and space in a few meaningful words. This provides us with very selected information but with the scanty data, we are able to reconstruct reality in compressed time and space. With this the text becomes a tyrant.

The Milieu-of-the-text

The Tyranny of the Text

The third quality of the text is worth a separate heading. Once the text is done, *the text breathes its own life*. A text is born. Your writing is dated when it was written or published. You put a date on your correspondence, a date is placed on a sourced reading material, a document is serialized with dates. But a text that is born also dies with its obsolescence. A text becomes a cadaver of obsolete content or inferior methodology. With a life of its own, it becomes an autonomous entity for *it creates a persona of its own*. It is a persona separate from the one who made it. Once a text is written, once it is born, it becomes another person with its own character and personality. Thus, a woman may fall in love with a man who writes sweet love letters, but she may not actually fall in love with the man, she fell in love with his letters, she fell in love with the text. The man who wrote it may have a different personality which she would eventually loathe but the love letters exist and may still generate the previous chill she experienced of the infatuation she may have felt. Even if the man disappears, the love letter could still survive for the letters had a life, a different persona from the detestable man who wrote it. She actually fell in love with the text which has a life and persona of its own.

This is evident in historical documents. Historical texts are repositories of human activities. The human activities that the texts capture present themselves as sources of historical analysis. The author of the text embodies the projection of the human subjects or even of himself or herself. But the text may persist even greater than what the author intended it to be to reveal about himself or herself or about the human activity that he or she is trying to capture. Louis Gottschalk has this to say:

> For the same reason the term personal document is the historian synonymous with the term human

> document. These terms were invented by social scientists. The historian is not likely to employ them. To him they appear tautalogous. All documents are both human and personal, since they are the work of human beings and she light upon their authors as well as upon the subjects the authors were trying to expound. Sometimes, indeed, they betray the author's personality, private thoughts, and social life more revealingly than they describe the things he had under observation. Here, too, a document's significance may have a greater relationship to the intention of the historian than to that of the author. Sometimes the historian may learn more about the author than the author intended that he should (Gottschalk 1969: 60-61).

If the text is a reservoir of human activity, Benedict Anderson assumed that nations are born with the text. Anderson mapped out a sociological definition of a nation as an imagined community, "*imagined* because the members of the even the smallest nation will never know most of their fellow-members, meet them, or even hear of them, yet in the minds of each lives the image of their communion" (Anderson 1991:6). This community born with national consciousness would not have emerged without the print as a commodity and technology.

> We can summarize the conclusions to be drawn from the argument thus far by saying that the convergence of capitalism and print technology on the fatal diversity of human language created the possibility of a new form of imagine community, which in its basic morphology set the stage for the modern nation. The potential stretch of these communities was inherently limited, and, at the same time, bore none but the most fortuitous relationship to existing political boundaries... Yet

> it is obvious that while today almost all modern self-conceived nations – and also nation-states – have 'national print-languages', many of them have these languages in common, and in others only a fraction of the population 'uses' the national language in conversation or on paper (Anderson 1991:46).

The text that breaths a life of its own is a reservoir of human activity and is an important element in creating a consciousness necessary for the imagination that would create a nation. Theories, however, are also texts. Theories once conceived and written also create a life of their own. The scientist dies but the theory still lives and takes its own persona. The theory is reviewed, it is used or it is attacked. In one writing or in a conference, you don't debate with the scientist anymore but you debate with the theory itself. You argue with the text. And most of the time, the theory answers back. The text has its retort. The limitation we try to defend the theory with considering the imperfections it incurs for some aspects of reality that it could not account for, pictures how the theory answers back as attacks are hurled against it. The theory as a bounded entity is born, lives and could die if it is proven to have lost its explanatory value. The text dies in obsolescence and irrelevance. Thus if humans can create theories which can survive on their own, humans can also kill them just as they can murder the text.

Our interaction with the text leads us to another form of reality which is ***textual reality***. Such interaction does not only include its perusal, but our critique of the text, our interpretation of it, our application, our extension of what it conveys comprises this reality. It does not matter at this moment, if the text conveys truth. That is a different matter though. But our interaction with this separate entity that lives in the words that we read, and which creates a persona of its own, is in itself an observable experience. It can even be counted how many times that experience was realized.

Elegant Convincing

If the text lives and breathes a life of its own and becomes its own persona, then what for should we present it in a scientific gathering? Theories though they become separate autonomous entities, are still constructed in the milieu-under-construction among humans as a matter of elegant convincing. Take for example Quantum Mechanics. There was the necessity for the Solvay Conference of 1927 to convene the best minds in physics not to find out if Quantum Mechanics works or to find its truth value for no amount of conferences could make it work or unwork if it really does or does not. Scientific conferences are a means to convince the community of scientists about a scientific finding or theory, which is a means to construct a text that has a life of its own in a milieu-under-construction.

The convincing is no ordinary persuasion. It is elegant. No amount of convincing could convince a community of highly learned individuals if it is not elegant. A theory would be debated and argued even further even without its original creator and its acceptance comes if it elegantly survives its beating. Thus, in the manner of social construction, the most illustrious among them in the community, the most respected of them elegantly wins the debate. In that sense, the high regard for the proponent ends up with the elegant acceptance.

Since the theory is a text made out of language, the text is also accepted if it were elegantly worded. Such elegance in its textual construction comes with how creative or beautiful it is worded or how difficult it could be read. Ockham's Razor tells us of a principle that economy of words is a criterion of a theory's acceptance. But we could also add Pythagoras' principle that beauty is a universe's quality. Similarly, social constructivity suggests that a theory's beautifully crafted language or the toughness of how it could be understood make for how the community of humans could accept the autonomous entity into their social circle. A scientist at one point should be a good writer... a wordsmith in capturing and reducing concepts. An artist is not the only creative person, a scientist also is... creative at spotting the uniqueness of a phenomenon, creative at designing the proper methodology and creative at expressing it into language.

Elegant convincing paves the way for debates among the proponents and opponents of the theory. But after the dust has settled, if the theory musters enough supporters, the text lives to assemble more disciples. This time, the disciples argue for the theory while the opponents argue against it. But they don't argue with the theorist anymore for the theorist may have long been dead. They debate for or still oppose the theory. But the arguments sometimes take a different turn, that as the disciples of the theory take hard line stance, the text has become pedestaled to appear absolute. Here is where the text becomes a tyrant. And the elegant convincing goes on.

Popper did not falsify nature. He falsified texts. The same is true with Lakatos who in the face of attacks against the text thought of core and peripheral assumptions to defend the text against attacks and adjust its assumptions. A paradigm which Kuhn coined is a text that had won ample disciples to use and defend. Here is where elegant convincing has become an integral part of the text and the enterprise of science itself.

The Textualized Nature of Theories

The *Textualized Nature of Theories*, therefore, assumes that:

- *The nature of the natural and social worlds suggests that they are both imbued with the properties of the milieu-already-constructed and the milieu-under-construction.*

This assumption veers away from dichotomizing the natural world to one that refers to the milieu-already-constructed and the social world to be that of the milieu-under-construction. The natural world has properties of it already made and being made while the social world is also imbued with one that is already constructed and one that is still being constructed. Perhaps, this is really how the natural and social worlds work. Biological life, for example, obeys certain laws. Genetics is law governed. The composition of the genes is definite and certain (milieu-already-constructed). But the process of how this genetic material is transmitted through the random unification of two cells in a meiotic combination is fraught with uncertainty (milieu-under-construction). But that is how life perpetuates. The law of mechanics works for objects with utmost certainty and predictability as to its speed (milieu-already-constructed) but at the atomic or quantum level things become crazy and uncertain (milieu-under-construction). But that is how light and even your television set works. Economics obeys the basic law of supply and demand as to the goods and price exchanged (milieu-already-constructed) but transactions and consensus made out of the meanings and perceptions of the product and the situation that brought it about is continuously being made everyday (milieu-under-construction). This is how the economy works. Nature and society then operates and perpetuates in these two milieus. This brings us to the next assumption that…

- *The phenomena arising from either milieu are multidimensional while the theories that are hoped to account for them are theme and problem specific.*

Theories are not conceived to compete with each other. They are crafted and boldly asserted in order to give account of the world which is actually multifaceted. A theory, therefore, could give account only of one face of the phenomenon while another theory could mirror two faces of it. This would explain what Lakatos is saying that one theory could explain more empirical evidences than another. But even if one theory could mirror more faces of the phenomenon while one theory is limited to explain only a single face of a phenomenon, it does not mean that one theory is inferior to the other, worse, we could not say that the perceived inferior theory is wrong just as Thomas Kuhn asserts. Each theory is specific of what it could give account of. It has its own limits. Beyond such boundaries, if pushed too far, the explanatory value of the theory breaks. It fails. Thus is Popper's falsification. If a theory works, then it does. If it were spurious, then no amount of falsification would do justice to it, because in the first place it is false.

Even the progress of science can be reckoned along this line. In order to give account of science and find a pattern of how science has progressed, theory also has been used to capture the pattern of such development with theories being the subject of such progress. The works of the philosophers of science like Popper, Kuhn and Lakatos are also themselves theoretical. The kernel of their works is theory and the pattern that arise are also encroached on the theories that they propound, leading us to the point that the practice of science is also multidimensional. The structure and dynamics of theory "building" are themselves theoretical and multidimensional.

- *Theories are text-independent entities that reflect and interpret two kinds of milieus: the milieu-already-constructed and the milieu-under-construction. This is the milieu-of-the-text.*

Text-independence characterizes theories as having an existence of their own. The text embodies the structure of theories as comprising of system-statements that provide the means of description, analysis and interpretation of the specific properties of the world whether it be of the milieu-already-constructed or the milieu-under-construction. The specificity of these system-statements imbues its capacity of closure or the creation of boundaries of one idea or concept of the world. By closure, the specific system-statements are able to encapsulate and provide description, analysis and interpretation of the specific properties of the world. Furthermore, the independent character of theories also characterize them as having a life of their own which embody the expressiveness of three realities: the sensed, axiomatic and interacted realities. Theories are able to interpret the observable or sensed properties of the world. Yet theories do not only embody a system of rules or they are themselves interpretative of rules or axioms but theories are governed by rules that make them exist on their own. Scientists and theorists interact with the world through theories. In like manner, they interact with theories by falsifying, corroborating, criticizing or defending them in various fora, creating an interacted reality. In fact the use of a theory to interact with the world is an agreement among members of the scientific community. Likewise, the more following a theory creates, the more illustrious scientists who would confirm or use the theory, the greater is the theory's acceptance among practitioners. Thus theories construct the world of scientists. And while theories reflect and interpret realities in the world, theories create a reality of their own. This is the milieu-of-the-text. And this is textual reality. Thus the milieu-already-constructed befits that of sensed and axiomatic realities; the milieu-under-construction suits that of interacted realities; while the milieu-of-the-text takes that of textual realities.

- *The deconstructive-reconstructive-foreconstructive capability is the heart of textualization enabling theories to capture and mirror the world.*

This capability separates theories from any other text. The capacity of theories to disassemble the properties of the world into parts or variables and to put them back again in order to account for their mechanics and dynamics is the central theme of textualization and the central characteristic of theories. Deconstruction-reconstruction-foreconstruction then is textualization and that equips theory with translating the world into text and making it readable. But the capturing of the world is not an all-embracing feat; rather, disturbances and anomalies come with the enterprise of textualization. Thus with the deconstructive, reconstructive and foreconstructive capabilities of theories, the mirroring of the world carries with it the textualization of regularities and irregularities, as well as disturbances that go with it. Thus textualizing the world is not a single effort to provide a complete and perfect picture of the world. The textualization of regularities, disturbances and anomalies provide only a certain reflection of an aspect of the world, thus one theory could hardly be judged better than the other. It would only depend on the choice of the scientist on what picture he would pick in order to look at the world and serve his aim at reconstructing it and his world, for every theory has its own limit or usability in deconstructing, reconstructing and foreconstructing the world, thus creating a certain picture of it.

- *Humans are the center of theories; humans are the central figures of science; theories construct and reconstruct their world and they construct and reconstruct the world with them.*

Theories are human inventions. Science also is a human invention. The world is meaningless by itself. But humans make sense of it by textualizing it. In the process of textualization, that is by deconstructing, reconstructing and foreconstructing an aspect of the world, the practitioner is engaged in finding patterns out of the set of regularities. But as practitioners pull out these patterns and weave them, irregularities are also exposed. In the same manner, the enterprise of scientific discovery, theory development, falsification and experimentation carry with them the disturbance of the system being textualized, deconstructed, reconstructed and foreconstructed. Thus it is not only the pattern of regularities that the theory accounts for that constructs and reconstructs the world of the scientist, theorist or any reader of it. But the world is also constructed and reconstructed with the irregularities exposed by the theory for as practitioners apply or attack the theory, they also know the forbidden boundaries that lie outside the regularities which the theory could not interpret. Thus as theories textualize the milieu-already-constructed or the milieu-under-construction with all the patterns of regularities and with the uncovering of anomalies together with the disturbances; the world of the practitioner is also built and rebuilt. In this regard, the practitioners could intervene and manipulate the world. It is "how we read the world and have come to change it."

In a milieu-already-constructed and milieu-under-construction, theories that are crafted thrive in the milieu-of-the-text that is constructed in the social world of humans. As the text autonomously takes life on its own, the theory as a separate entity from the object of milieu-already-constructed or milieu-under-construction under study, carves its own place in the community of humans with its elegant convincing. The theory lives and outlives its maker just as the text survives through time further enabling humans to "read the world and change it."

- *The text has power, theories possess power; textualization is power.*

Theories as texts have power because they can subject the world under scrutiny. Power here, that theories possess does not concern the empirical content where we could say one theory is more powerful than the other because it has more falsifiable statements. That is for Popper to defend. Power here does not also mean the capability to intervene or to manipulate. Joseph Rouse, in his book *Knowledge and Power: Toward a Political Philosophy of Science,* postulates science as a "way of acting on the world, rather than a way of observing and describing it" (Rouse 1987: 129). He subscribes to the Foucaultian concept of power in terms of gaze and surveillance (Foucault 1995) that seethes inside and outside the laboratory. Power for Rouse is in the practice of doing science, "from representation to manipulation… from knowing than to knowing how… that science helps disclose the world" (Rouse 1987: 25). Power then in the present project adheres to the concept of the deconstructive-reconstructive-foreconstructive capability of theories to put the milieu-already-constructed or the milieu-under-construction under scrutiny by breaking them into parts and assembling them back again. But that is not the end of it. The power that theories possess does not only pertain to its deconstructive-reconstructive-foreconstructive capability but also to its capacity to build and rebuild the world of human beings. The construction and reconstruction of the world bring to fore the concept of power or what this project calls "self-empowering power." By self-empowering it means its deconstructive-reconstructive-foreconstructive property. It is the ability to bring about, to create and recreate. It is like a device which when built, builds and rebuilds other things. It is like a tool which when fashioned devices other objects. And it is accomplished in the milieu-of-the-text.

Postscript 3

How does the *Textualized Nature of Theories* provide the middle ground between the knights of science and the postmodern assaults?

Table 6.5 Textualized Nature of Theories' Middle Ground

Science	Postmodernity's Attack	Textualized Nature of Theory
One of the goals is the unification of theories into a grand theory.	Despises grand theories and universals as it is skeptical of metanarratives.	While not one metanarrative could explain everything in nature and in the social world, the grand theories that science formulates have the capability to explain some overarching principles of nature and social order in the milieu-already-constructed or milieu-under-construction.
Achieves objectivity using theory and rigid methodology.	Scientists are still part of the social constructions and are not free from ideological bias.	While scientists are part of a community and are affected by biases (this is the milieu-under-construction at work), they could

Science	Postmodernity's Attack	Textualized Nature of Theory
		also find detachment to achieve objectivity (this is milieu-already-constructed in operation) through theories.
Language is a tool to convey scientific findings.	Language is a tool of persuasion and power.	Theories are texts that thrives in the milieu-of-the-text where language is not just a tool but also a means of persuasion for elegant convincing.

While a metanarrative to explain everything as the postmodernists are skeptical about, a grandest of the grand theory to explain everything may not really be feasible. Nature and social relations are multifaceted to be accounted for by one grandest of grand theories. But postmodernist's incredulity to a metanarrative and a claim that there is no such metanarrative is a metanarrative in itself. Science has made success in uniting theories. Maxwell's Electromagnetic Theory combines electricity and magnetism, the Special Theory of Relativity unites time and space - are examples of these successes. Grand theories or unified theories are still a slice of the whole immense magnitude of nature and social relations. The grand or unified theories that science has devised has struck success because of the overarching principles operating in the natural and social worlds as well.

Objectivity, which is science's ultimate nature is in danger of being disproved due to postmodernity's claim that a scientist is a member of the society and he or she cannot be completely withdrawn from biases. Science has become a culture in the postmodern era where the judge of acceptable knowledge is that which is categorized as scientific and nothing else. This, however, is part of the milieu-under-construction where the scientist constructs and reconstructs science into his or her social world. But there still exists a world which will operate even if science discovers it or not, although, science can only investigate it once the scientist disturbs it. This, however, is the milieu-already-constructed. As previously discussed, Quantum Philosophy asserts that the complete detachment of the scientist from the object of observation is impossible to attain since any scientific endeavor necessitates its disturbance of the object itself. Quantum Philosophy suggests that such disturbance results in the observer becoming part of the observed. Here, absolute objectivity would never be attained, lest the scientist does nothing and does no science as a result. Yet this does not doom science for the only means of detachment that a scientist does is the use of instrument and methodology characterized with rigor. This is the only way were objectivity is attained and this is instrumental objectivity as previously discussed.

Lastly, language is a medium where the discovered milieu-already-constructed and the milieu-under-construction could perpetuate. While language is a tool, it is also a repository of the beauty or toughness of language where the discovered and constructed can be relayed, understood and accepted... and it is most of the time perpetuated because of the how elegantly it was able to convince. The tool has power. The text which perpetuates with the life of its own has power. It is able to convince and the elegance of how it convinced is in the text itself. This is the milieu-of-the-text.

A theory is a textualized system-formulations of the deconstructed, reconstructed and foreconstructed phenomenon. A phenomenon, in this context, can be defined as a regularized or patterned occurrence. A theory is a set of statements meant to formulate concepts that attempt to mirror phenomena that besiege human beings in the natural or social order, where such statements are capable of deconstructing the phenomena into parts, reconstructing or building them

The Milieu-of-the-text

back again into a relatedly functioning whole, and capable of providing foreknowledge of how such relatedly functioning whole operates. Science is rigor-bound, method-driven, textualization of patterns of regularities and irregularities in phenomena that exist in nature or in human relations through deconstruction, reconstruction and foreconstruction in an elegantly convincing fashion. Science is not just nature or human relations textualized, it is also the method and validation textualized. Science then is a body of rigor-achieved, rigor-differentiated and rigor-validated text that can create an image of the world it attempts to explain or understand; represent or model; discover its order or even re-order what it has discovered. Science is a human invention, inasmuch as theories also are.

The Semantic View of Theories "identifies theories with certain kinds of abstract theory-structures, such as configured state spaces, standing in mapping relations to phenomena. Theory structures and phenomena are referents of linguistic theory-formulations" (Suppe 2000:105). The Semantic View of Theories assumes theories that frame, mirror, model or map phenomenon. The Textualized Nature of Theories presuppose that while theories are composed of statements, these texts are able to reflect the world that is already constructed or under construction.

Theories then are interpretations of the worlds, borne out of textualization, built out of the patterns set by regularities, which expose anomalies in return. And while incorporating the disturbances in the system being studied, theories structure and restructure, build and rebuild the world of human beings, empowering them to manipulate the worlds that were then structured. Power, therefore, that theories possess is self-empowering. If humans invent theories, theories invent them as well.

This is the beauty that we have come to understand these worlds. This is the wonder of the worlds that we have come to read them. And through the way we read them proceeds the means on how we could change them. And the more we want to read and change them, the more we try to invent and re-invent the means to textualize them.

The natural, biological and social worlds are multi-faceted. The more we try to unlock their nature, the more patterns we discover, the more relationships we uncover but the more anomalies we also expose and the more limits of our textualization we set. As we see more limits, the more we desire to investigate and the more we search the more we end up starting at the beginning.

> *For in the beginning was the word, and the word was born out of reason, so in the beginning was the text, and this text was theory.*

Chapter 7

...A Way to Begin Again

Brussels, Belgium... 1927... after almost a week of grueling intellectual exchanges, the 5th Solvay Conference ended with the victory of the adherents of Quantum Mechanics. What transpired was a debate that elegantly convinced them except for Einstein who was convinced but not elegantly and stated that Quantum Theory was incomplete. The constructive nature of the minute will operate no matter how much debating they had engaged in but they were able to convince each other and the text they produced eventually convinced us as well. It was done elegantly and we are convinced elegantly as well. The constructive nature of the minute was then constituted into our social world; into our knowing and into our being convinced and eventually into our manipulating the minute using the theory. Our civilization has only two things now: us and the text. If our civilization ends, what will be left are our bones and the text. Even the artifacts will find meaning in the text that we interpret the artifacts to be. This is how science proceeds.

Reading the World and Inventing Science

If we look up the heavens, watch stones fall to the ground, hear the waves slap on the shore, peer at the cell into the microscope, we don't see science. What we see is nature. When we look for patterns how nature operates, that is science. If we observe how people behave or synthesize what they write or say, we don't witness science, we experience human interaction. But when we look for patterns in human behavior, then what we are witnessing is science. Our search for patterns in order to find out how nature operates and how human interactions proceed and are reproduced will not materialize without putting them into text. The text is our invention. Science is our invention.

In this invention we have also come to realize the nature of things and also of ourselves. Nature itself is constituted with both the determined and the probabilistic, the certain and the uncertain, the pre-fashioned milieu-already-constructed and the reconstitutable milieu-under-construction which we perpetuate and reconstitute through the milieu-of-the-text. In the text, science survives. This is our invention.

What then do philosophers of science do within this scientific enterprise?

Philosophers of science bother themselves with the systematics of knowledge. Science is knowledge epitomized. Philosophers of science ask questions of how scientific knowledge proceeds and is reproduced. Kuhn says it is by revolution. Popper admits it is methodical by proving the theory wrong rather than proving it right. Lakatos debates it is through the adjustments of the main and peripheral arguments.

Besides the bothering themselves with the systematics of knowledge, philosophers of science entangle themselves with the issues of nature and human interaction. Philosophers of science ask questions which they cannot answer or hazard an answer which they cannot prove. But out of these attempts nature is textualized. These texts are theories which scientists challenge to falsify through their enterprise whether by experimentation or observation, simulation or mathematical formulation. Thus, philosophers of science throw the questions or attempts to respond to these questions as scientists come with their elegant convincing to finally attempt to strike the gavel and resolve the issues. In the process, knowledge grows. And surely, science is nothing but knowledge or episteme.

Aristotle asked which motion is superior and how do planets move? His answer... circular motion is superior because it has no end unlike the rectilinear one and since planets are situated in the heavens where the gods are, then the planets travel in a circular path because it is up there in the heavens close to the gods.

Aristotle attempted an answer to his problematique but it had to wait for its approval. It had to wait for Galileo to invent his telescope and peer into Jupiter in order to prove the circular movement of the planets as he saw the moons of Jupiter orbit around the giant planet. Through logical deduction, he concluded that it is the earth that orbits around the sun and not otherwise, though he never saw the earth doing the orbiting in his invented eyepiece. It took Newton to formulate how much is the attraction of the bigger body in space to the smaller one. It took Einstein to draw a picture that this attraction is an indention on the fabric of space united with time.

But Aristotle's problematique and his response became the foundations of theories that present for their challenging by falsifying them in order to prove him wrong. We could not have arrived at how we view the world and the universe if all these articulations were never placed in the text. And everything that we know of science is in the text and science is a text that has its own life although the personalities who have elegantly convinced us are already dead.

The never ending search for patterns in the natural and social worlds will find life in the text. The mastery of the readable world is in the theory that we devised and these great events in science have found perpetuity in the text that we have also written. This is where the knowing knows no boundaries and the beginning always begins anew.

Figure 7.1 Group Picture of the 5th Solvay Conference Attendees, Brussels, Belgium, 1927

Front row from left: Irving Langmuir, Max Planck, Marie Curie, Hendrik Lorentz, Albert Einstein, Paul Langevin, Charles-Eugène Guye, C.T.R Wilson, Owen Richardson. **Middle row from left**: Peter Debye, Martin Knudsen, William Lawrence Bragg, Hendrik Anthony Kramers, Paul Dirac, Arthur Compton, Louis de Broglie, Max Born, Niels Bohr. **Back row from left**: Auguste Piccard, Émile Henriot, Paul Ehrenfest, Édouard Herzen, Théophile de Donder, Erwin Schrödinger, JE Verschaffelt, Wolfgang Pauli, Werner Heisenberg, Ralph Fowler, Léon Brillouin.

_____oOo_____

References

of Science, Robert Klee (ed.) New York: Oxford University Press.

_____ (2000) "Understanding Scientific Theories: An Assessment of Developments, 1969-1998," *Philosophy of Science*, Vol. 67, September, pp. S102-S115 (http://www.jstor.org/stabe/188661).

Taylor, Charles (1977) "Interpretation and the Science of Man" in Rabinow, Paul and Sullivan, William, (ed.) *Interpretative Social Science*. California: University of California Press.

_____ (1985) *Human Agency: Philosophical Papers I*. Cambridge: Cambridge University Press.

Toomer, G.J. (1975) "Ptolemy," *Dictionary of Scientific Biography*, Vol. 11, Charles Coulstan Gillispie (ed.) New York: Scribner and Sons.

Uzgalis, William (2012) "John Locke," *Stanford Encyclopedia of Philosophy*. http://plato.stanford.edu/entries/locke/.

Von Fritz, Kurt (1975) "Pythagoras of Samos," *Dictionary of Scientific Biography,* Vol. 11, Charles Coulstan Gillispie (ed.) New York: Scribner and Sons.

Weaver, Robert F. and Hedrick, Philip W. (1991) *Basic Genetics*. New York: Wm C. Brown Publishers.

Weber, Max (1968) Parliament and Government in a Reconstructed Germany, in G. Roth and C. Wittich Eds. *Economy and Society*, Tubingen: J.C.B. Mohr as quoted in Kalberg, Stephen (2001) The Modern World as a Monolithic Iron Cage? Utilizing Max Weber to Define the Internal Dynamics of the American Political Culture Today, *Max Weber Studies* 178-195.

Wilson, Robert (1997) *Astronomy through the Ages*, London: Taylor and Francis.

Winkler, Kenneth (ed.) (1996) *John Locke: An Essay Concerning Human Understanding,* Indianapolis: Hackett Publishing Company, Inc.

Index

A

a posteriori knowledge, 138, 261
a priori knowledge, 138, 261
acceleration, 71
Adams, John Couch, 57
adenine, 98-99
Adorno, Theodor, 195, 209, 266
Age of Enlightenment, 55-56
agreement, 138
Allegory of the Cave, 60,165
alleles, 91
amino acids, 90, 103
anaphase, 162 (See also mitosis)
anaphase, double, 1162 (See also meiosis)
Anaximanes, 136
Anderson, Benedict, 317
animalcules, 90
anomie, 265
anti-realism, 228
Archimedes, 21-22
Aristotle, 10-18, 69, 260

atoms, 57
Avery, Oswold, 97

B

bacteria ϕX174, 160
Benjamin, Walter, 266
Berger, Peter, 166
Berlin Observatory, 57
Berlin Physical Society, 146
biological systems, 89
Blumer, Herbert, 169
Bohr, Neils, 1
Boveri, Theodor, 91

C

capitalism, 192
Carnap, Rudolph, 270
cell division, 158, 223 (See mitosis and meiosis)
cell membrane, 90
Cell Theory, 90
cell wall, 90
cell, 90-91
centromere, 159, 162
Chamberlin, Edward, 120

Chargaff, Erwin, 98
chromatid, 159
Chromatin, 159
Chromosome Theory of Inheritance, 91, 157, 163
chromosomes
  colored bodies, 94-95
  crossover, 164
  X and Y 92
classical conditioning, 184
codon, 103
communicative action, 210-212
communicative rationality, 210-214
communism, 192
comprehensibility, 212
consensus, 212
constructivism, 229
Copernican Theory, 2, 29-32, 260
Copernicus, Nicolaus, 29
Correspondence Theory of Truth, 275
Crick, Francis, 99
crossover, 164
currency, local and foreign, 125-127
curvature (See Relativity, General Theory of)
cytoplasm, 90
cytosine, 98

D

Darwin, Charles, 273, 306
de Broglie, Louis, 149
deconstruction, 306

demand curve, 114 (See also Law of Supply and Demand)
Democritus, 136
deoxyribonuclease, 98
deoxyribose, 98
Descartes, Rene, 42-43, 138
Dialectic-Historical-Materialism, 192-193
dialectics, 189
distance, 70
disturbance (in science), 228-233
DNA (deoxyribonucleic acid), 98-100, 102-103
dominant traits, 152
double-split experiment, 226
Drosophila melanogaster, 91, 161
Durkheim, Emile, 265

E

economics, 104-106
EDSA-QA, 171-172
Efficient Market Hypothesis, 203
Einstein, Albert, 1-2, 74
Einsteinian mechanics, 70
electromagnetic force,
electromagnetic waves, 145
electron, 149-147
electrophoresis, 98
elegant convincing, 54, 319
empiricism, 137
endoplasmic reticulum, 91
energy (from mass), 89
Enlightenment, 60
epicycle, 24-25

Index

equilibrium (of price and quantity), 113
Eratosthenes, 16
Escherichia coli (*E. coli*), 160
ether, 81
Euclid, 20
Euclidean geometry, 252-255
exchange rates, 127
experiment, 231-232

F

falsification, 276-280
Fama, Eugene, 203
feudalism, 192
Feurbach, Ludwig, 190
Fiegl, Herbert, 270
flatlander, 77-78
flatworld, 77
force, 71
Foucault, Michel, 304
fourth dimension, 87, 256
Franklin, Rosalind, 99
Fromm, Eric, 266

G

Galile, Galileo, 2, 35-38, 73, 222
Galle, Johanne, 58
Game Theory, 197-201
gametes, 92, 161
genes, 91-92
genetic code, 103
genotype, 95
Geo-centric theory, 13-15, 23, 260
geometry, 252
Giddens, Anthony, 300
Gödel, Kurt, 270
Gottschalk, Louis, 317
gravitational constant, 248
gravity, 249-259
Grew, Nehemiah, 89
Griffith, Frederich, 96
Grounded Theory, 168-169
guanine, 98

H

Habermas, Jürgen, 209, 266, 268
Hawking, Stephen, 309
Hegel, Georg Frederick, 189
Heisenberg, Werner, 148
helio-centric universe, 37, 260
Hempel, Carl, 276
Heraclites, 136
hereditary traits, 165 (See also genes, genotype, phenotype)
heredity, 156
Herschel, William, 57
heterozygous, 95
Hicks, John, 116
homozygous, 95
Hook, Robert, 189
Horkheimer, Max, 195, 209, 266
Hubble, Edwin, 222
human will, 183
Hume, David, 45-46

I

Indifference Curve Analysis, 116-117
inertial frame, 71-76
Institute of Social Science Research, Frankfurt, 209

instrumental, strategic action, 210
interest rate, 126
interphase, 157 (See also mitosis, meiosis)
interpretative, creative statements, 243
intersubjective
   meaning, 171
   rules, 78
   world, 166
invasion
Iron Cage, 209, 264
irregularities, 245
   natural, 235

J

Jakob, Matthias, 90
Janssen, Hans, 89
Janssen, Zacharias, 89

K

Kant, Emmanuel, 138, 242, 261
karyotype, 92
Kaufman, Felix 270
Kepler, Johannes, 32-34
Kepler's Planetary Motion, 33-34, 245, 250
kinetic energy, 85
knowledge, 137-138
Kornberg A., 160
Kraft, Victor, 270
Kroto, Harold, 51
Kuhn, Thomas, 280-283, 295-299

L

Labor Theory of Value, 191
Lakatos, Imre, 288-295, 296-299
Law of Diminishing Marginal Utility, 115
Law of Dominance, 152
Law of Independent Assortment, 154
Law of One Price, 124
Law of Purchasing Power Parity, 124
Law of Segregation, 152
Law of Supply and Demand, 110 (See also price)
   demand curve, 112, 114
   equilibrium, 112
   fluctuation in exchange rate, 127
   mechanics, 110
   supply curve, 112, 114
Leeuwenhoek, Anton Van, 89
Leverrier, Urbain Jean, 58
lifeworld, 213
light (See speed of light, quantum theory)
Logic of Scientific Discovery, 276
logic, 244
Logical Empiricism, 271
Logical Positivism, 270
Lorentz, Hendrik, 1, 86
Lowenthal, Leo, 266
Luckman Thomas, 166
Lyotard, Francois, 301

M

Macleod, Colin, 97

Index

Malpighi, Marcello, 89
Malthus, Thomas, 111
Marcuse, Herbert, 195, 209
Marcuse, Herbert, 266
market, 113
Marshal, Alfred, 112
Marx, Karl, 188
mass, 71
mathematics, 241
matter, 74
Maxwell, James Clerk, 80, 145
McCarthy, Maclyn, 97
meanings, 170
measurement, 240
mechanics, Einstenian, 143-144 (See also relativity)
mechanics, Newtonian , 142 (See also Law of Motion)
meiosis, 158-162
Meischer, Friedrich, 96
Mendel, Gregor, 92, 151
mercantilism, 107
messenter RNA (mRNA), 102
metaphase, 158
metaphase, double, 162
methionine, 103
Michelson, Michael, 81
Michelson-Morley experiment, 246
Milieu-already-constructed 57-63, 223-228
Milieu-of-the-text, 65, 324
Milieu-under-construction, 58-61, 223-228
Mill, John Stuart, 111
mithocondria, 91
mitosis, 157
Modernity, 262
Molecular Theory, 157

Morgan, Hunt, 91
Morley, Edward, 81
Mun, Thomas, 108

N

Neptune, 257
Neurath, Otto, 270
Newton, Isaac, 70
Newtonian physics, 69, 70-72
Nirenberg, M.W., 103
No Boundary Proposal, 227, 311
normal science, 281
nucleus, 91
number, 242

O

observables, 246
Ockham, William of, 27-28
Ockham's razor, 28
optimist, quantum disciples, 226
organelles, 90

P

paradigms, 281-283
particle-wave, 149 (See also quantum mechanics)
Pauling, Linus, 89
Pavlov, Ivan, 184
pessimists, quantum disciples, 226
phenotype, 94
phenylalanine, 103
photon, 147 (See also quantum theory)
Planck, Max, 145

Planck's constant, 146
Plato, 55, 165
Pollock, Friedrich, 209, 266
Popper, Karl, 276280, 295-299
Postmodernity, 300-306
    attack on science, 307-315
power, 305
pragmatist, quantum disciples, 227
praxis, 10
price, 110-124
Prisoner's Dilemma, 199-201
prokaryotic cells, 90
proletariat, 192
prophase, 158
prophase, double, 162
proteins, 101
Protenor, 92
Ptolemy, Claudius, 18
Punnett square, 93
purines, 99
pyrimidines, 99
Pythagoras, 19-20
Pythagorean Theorem, 79

Q

quanta, 147
quantum disciples, 223
quantum mechanics, 114, 150, 223-227, 228
quantum theory, 146

R

radiation, 145
rationalist, 137, 165
realism, 228
reality, 137-139, 227
Received View, 50, 270

recessive trait, 152
recursive-interactive acts, 186
Reimann, Bernhard, 252
Reimannian geometery, 257-259
Relativity, General Theory of, 1, 258-259, 287
Relativity, Special Theory of, 74-87, 290
replication, 100
ribosomes, 91
Ricardo, David, 109
rightness, 211-212 (See also validity claims)
RNA (ribonucleic acid), 98, 100
Robinson, Joan, 120
Rouse, Joseph, 326
Rousseau, Jean Jacques, 261
Rubens, Heinrich, 145

S

satisfaction (See utils, utility)
Schrodinger, Erwin, 149
Schrodinger's cat, 225
science, 50-54
scientific community, 281
scientific revolution, 280-283
self-empowering power, 327
self-interpreting beings, 166
self-inventing beings, 183, 186-188
sincerity, 212 (See also validity claims)
Smith, Adam, 108-109
socialism, 192
space, 75
space-time, 87, 143, 252, 259

Index

speech acts, 211-213
speed, 70, 87
speed of light, 84, 87-88
Strand Separation Theory, 160
Streptococcus pneumoniae, 96
supply, 111
supply curve, 112 (See also Law of Supply and Demand)
Sutton, Walter, 91
symbolic interaction, 167

T

telophase, 159 (See also mitosis)
telophase, double, 162 (See also meiosis)
Textualilzed Nature of Theories, 6-7, 321-327
textualization, 83, 330-331
Thales, 136
theoria, 10
Theory of Communicative Action, 210-214
Theory of Heredity, 156
Theory of Immanent Form, 137
Theory of Motion, 18
Theory of Natural Selection, 273
Theory of Perfect Competition, 118-119
Theory of Photoelectric Effect, 224
Theory of Scientific Growth, 291
Theory of Time, Information and Money, 121

Theory of Transcendental Form, 136
thesis-anti-thesis-synthesis (See dialectics)
Three-dimentional world, 79, 86
thymine, 98
time, 88
traffic, 170
transfer RNA (tRNA), 103
truth (objective), 212 (See also validity claims)
Two-dimentional space, 75-77

U

ultracentrifugation, 98
Ultraviolet Absorption Spectrophotoresis, 98
Uncertainty Principle, 148, 224 (See also quantum mechanics)
Universal Theory of Gravitation, 287
uracil, 98
Uranus, 57
utility, 114
utils, 114

V

validity claims, 211-212
velocity, 88 (See also mechanics, speed)
Vienna School, 270
Virchow, Rudolf, 90

W

Watson, James, 99

wave mechanics, 224
Weber, Max, 209, 263
Wilkin, Maurice, 99
world-line, 143-144

X
Xenophanes, 136
X-ray diffraction, 99

Z
zero-sum game, 199

www.ingramcontent.com/pod-product-compliance
Lightning Source LLC
Chambersburg PA
CBHW031606210526
45464CB00004B/1448

References

Adler, Mortimer J. (ed.) (1993a) "Johannes Kepler: Epitome of Copernican Astronomy" translated by Charles Glenn Wallis, *Great Books of the Western World*, Vol. 16, Chicago: Encyclopedia Britanica, Inc.

_____ (1993b) "Nicolaus Copernicus: On the Revolutions of the Heavenly Spheres" translated by Charles Glenn Wallis, Great *Books of the Western World*, Vol. 16, Chicago: Encyclopedia Britanica, Inc.

_____ (1993c) "Ptolemy: The almagest" translated by R. Catesby Tliaferro, *Great Books of the Western World*, Vol. 16, Chicago: Encyclopedia Britanica, Inc.

Anderson, Benedict (1991) *Imagined Communities: Reflections on the Origin and Spread of Nationalism*. London: Verso.

Ariew, Roger, (ed.) (2000) *Rene Descartes: Philosophical Essays*, Indianapolis: Hackett Publishing Company, Inc.

Atkinson, Rita L. Atkinson, Richard C. and Hilgard, Ernest R. (1983) *Introduction to Psychology*. New York: Harcourt Brace Jovanovich, Inc.

Ball, Terence (1987) "Is there Progress in Political Science?" in *Idioms of Inquiry*. Ball, Terence (ed.) Albany: State University of New York Press.

Beenaker, Carlo (2015) *Lorentz and the Solvay Conferences* https://www.lorentz.leidenuniv.nl/history/Solvay/solvay.html

Berger, Peter and Luckman, Thomas (1993) "Foundation of Knowledge in Everyday Life." in Farganis, James (ed.) *Readings in Social Theory*. New York: McGraw Hill.

Blumer, Herbert (1993) "Society as Symbolic Interaction." in Farganis, James (ed.) *Readings in Social Theory*. New York: McGraw Hill.

Brealey, Richard A., Myers, Stewart C., and Marcus, Alan J. (2004) *Fundamentals of Corporate Finance*. Boston: McGraw-Hill.

Bulmer-Thomas, Ivor (1971) "Euclid," *Dictionary of Scientific Biography*, Vol. 4, Charles Coulstan Gillispie (ed.) New York: Scribner and Sons.

Burns, George W. (1972) *The Science of Genetics: An Introduction to Heredity*. 2nd ed. New York: McMillan Co.

Business Insider, Science (2015) http://www.businessinsider.com/solvay-conference-1927-2015-4

Carls, Paul (2016) "Emile Durkheim," *Internet Encyclopedia of Philosophy*, (http://www.iep.utm.edu/durkheim/)

Carnegie Science (2018) Carnegiescience.edu "1610 observing the moons of Jupiter" (https://cosmology.carnegiescience.edu/timeline/1610/observing-the-moons-of-jupiter)

Choi, Charles Q. (2015) "Different Tastes: How Our Human Ancestors' Diet Evolved," *LiveScience*, January 16 (http://www.livescience.com/49477-human-ancestors-diet-taste-evolution.html)

References

Clagett, Marshal (1970) "Archimedes," *Dictionary of Scientific Biography,* Vol. 1, Charles Coulstan Gillispie (ed.) New York: Scribner and Sons.

Cole, Nicki Lisa (2015) "Understanding Max Weber's Iron Cage" *Sociology About.* (http://sociology.about.com/od/Key-Theoretical-Concepts/fl/Understanding-Max-Webers-Iron-Cage.htm)

Copleston, Frederick (1960) *A History of Philosophy: From the French Enlightenment to Kant,* Vol. VI, New York: Image Books.

Corradeti, Carlo (2016) *The Frankfurt School and Critical Theory European Academy and the University of Rome* (http://www.iep.utm.edu/frankfur/)

Curtis, Michael, ed. (1981) *The Great Political Theories,* Vol. 1, New York: Avon Books.

Downing, Lisa (2011) "George Berkeley" *Stanford Encyclopedia of Philosophy*, (http://plato.stanford.edu/entries/berkeley/).

Durant, Will (1961) *The Story of Philosophy*, New York: Washington Square Press .

Eaton, Curtis, Eaton, Diane and Allen, Douglas (2010) *The Theory of Perfect Competition,* Pearson Microeconomics (http://wps.prenhall.com/ca_ph_eaton_microecon_6/22/585 1/1497893.cw/index.html)

Fama, Eugene F. (1995) Random Walks in Stock Market Prices *Financial Analysts Journal*, September/October 1965 reprinted January-February 1995 (http://www.investorhome.com/emh.htm Accessed 2 July 2009)

Farganis, James (ed.) (1993) *Readings in Social Theory*. New York: McGraw Hill.

Ford, Kenneth W. (1974) *Classical and Modern Physics*. Vol. 3. New York: John Wiley and Sons.

Foucault, Michel (1975) *Discipline and Punish: The Birth of the Prison*, Trans. By: Sheridan, Alan, New York: Vintage Books.

Foucault, Michel (1995) *Discipline and Punish*. New York: Vintage Books

Fromm, Erich (1965) *Escape from Freedom*, New York: Avon Books.

Gabriel, Percival S. (2012) News and money in times of political instability: Measuring the expressivity of newspaper articles and its effect on the peso-dollar exchange rate and the Philippine Stock Market index. *International Academic Research Journal of Economics and Finance*. 1(3): 36-50.

_____ (2013) How Newspaper-article-events, Other Stock Market Indices, and the Foreign Currency Rate Affect the Philippine Stock Market. *Asian Economic and Financial Review*. 3(4): 423-444.

_____ (2014) The Theory of Time, Information and Money in a Competitive Market, *Economics,* Vol. 3, No. 1: 9-18, doi:10.11648/2014301.12.

Gale Document (2013) *Standards of Beauty are Determined by Evolutionary Biology* (http://ic.galegroup.com/ic/ovic/ViewpointsDetailsPage/DocumentToolsPortletWindow?displayGroupName=Viewpoints&u=nysl_ro_bri&p=OVIC&action=2&catId=&documentId=GALE%7CEJ3010659230&zid=8444ff8f06932fd7686ca92853bbf96d)

"Galileo and the Inquisition" (2015) (http://physics.ucr.edu/~wudka/Physics7/Notes_www/node52.html)

References

Gamow, George (1961) *One Two Three... Infinity*. New York: Bantam Books.

Gibbons, Ann (2013) "The Evolution of Diet," *National Geographic*. February (http://www.nationalgeographic.com/foodfeatures/evolution-of-diet/)

Giddens, Anthony (1984) *The Constitution of Society: Outline of the Theory of Structuration*. Cambridge, New England: Basil Blackwell.

_____ (1990) *The Consequences of Modernity*. California: Stanford University Press.

Glaser, B. and Strauss, A. (1967) *The Discovery of Grounded Theory: Strategies for Qualitative Research*. Chicago: Aldine.

Gottschalk, Louis (1969*) Understanding History: A Primer of Historical Method*. New York: Alfed A. Knopf.

Habermas, Jurgen (1984) "Theory of Communicative Action" *Reason and Rationalization of Society* Vol. 1. Boston:Beacon.

Hawking, Stephen (1988) *A Brief History of Time*. New York: Bantam Books.

Hempel , Carl (1999) A Critique of Operationalism, in *Scientific Inquiry*, in Robert Klee, (ed.) New York: Oxford University Press.

Holt-Jensen, Arild (1999) *Geography History and Concepts,* London: Sage Publications.

Horgan, John (2003) *Quantum Philosophy*. [online] Cited March 13, 2003. (http://www.fortunecity.com/emachines/3ll/86/qphil.html)

Hume, David (2000) "Book 1 of the Understanding in Norton", David F. and Norton, Mary J. (eds.) *David Hume: A Treatise of Human Nature*, Oxford: Oxford University Press.

Hutchins, Robert M. (ed.) (1952) "Bacon, Francis: Novum Organum," *Great Books of the Western World*, Vol. 30, Chicago: Encyclopaedia Britannica.

Hutchins, Robert M. (ed.) (1952a) "Aristotle: On the Heavens," translated by W.D. Ross, *Great Books of the Western World*, Vol. 9, Chicago: Encyclopaedia Britannica.

_____ (1952b) "Aristotle: Posterior Analystics," translated by W.D. Ross, *Great Books of the Western World*, Vol. 9, Chicago: Encyclopaedia Britannica.

_____ (1952c) "Aristotle: Physics," translated by W.D. Ross, *Great Books of the Western World*, Vol. 9, Chicago: Encyclopaedia Britannica.

Huxley, G.L. (1975) "Eudoxus of Cnidus," *Dictionary of Scientific Biography*, Vol. 4, Charles Coulstan Gillispie (ed.) New York: Scribner and Sons.

Ison, Stephen (1993) *Economics*. London: Pitman Publishing.

Jocano, Landa F. (1975) *Philippine Prehistory: An Anthropological Overview of the Beginnings of Filipino Society and Culture.* Quezon City: Philippine Center for Advanced Studies, University of the Philippines System

Kaku, Michio (1994) *Hyperspace*. New York: Anchor Books.

Kaku, Michio and Thompson, Jennifer (1987) *Beyond Einstein.* New York: Anchor Books.

Kramer, Melody (2013) "The Physics Behind Schrodinger's Paradox," National Geographic (http://news.nationalgeographic.com/news/2013/08/130912

References

-physics-Schrodinger-erwin-google-doodle-cat-paradox-science/)

Kreis, Steven (2012) The History Guide: Lectures on Modern European Intellectual History. Plato Allegory of the Cave (http://wwww.historyguide.org/intllect/allegory.html).

Kroto, Harold (2008) *There are two type of Theories – Scientific and Unscientific Theories* (http://thesciencenetwork.org/docs/BB3/Kroto_Theories.pdf)

Kuhn, Thomas (1970) *The Structure of Scientific Revolutions*. Chicago: University of Chicago Press.

_____(1972) "Scientific Paradigms" in *Sociology of Science*, Barnes, Barry ed. England: Penguin Books.

Lakatos, Imre (1968) "Criticism and the Methodology of Scientific Research Programmes" Meeting of the Aristotelian Society at 21, Bradford Square London, W.C.I on the 28th of October. 7:30 pm.

Leffel, Jim (2000) Science and Postmodern Criticism. *The Scientific Review of Alternative Medicine*. Vol. 4, No. 1. 50-55.

Losee, John (2001) *A Historical Introduction to the Philosophy of Science*, New York: Oxford University Press.

Lyotard, Jean -Francois (1979) *The Postmodern Condition: A Report on Knowledge*. Trans. By: Benningnton, Geoff and Massumi, Brian. Manchester: Manchester University Press.

Mankiw, Gregory N. (1998) *Principles of Economics*. New York: The Dryden Press.

Marshall, Alfred (1890) *Principles of Economics*. London: MacMillan.

Marx, Karl and Engels Friedrich (1964) *The Communist Manifesto*. Moore, Samuel (trans.) and Katz, Joseph (ed.) New York: Washington Square Press.

Marx, Karl. (1981) "Introduction to the Critique of Political Economy." in Curtis, Michael (ed.) *The Great Political Theories*, Vol 2. New York: Avon Books.

Maxson, Linda R. and Daugherty, Charles H. (1989) *Genetics: Human Perspective*. Iowa: W. C. Brown Publishers.

Merzbach, Uta C. and Boyer, Carl B. (2011) *A History of Mathematics*, New Jersey: John Wiley and Sons.

Michon, Gerard P. (2015) *Solvay Conferences* (http://www.numericana.com/fame/solvay.htm)

Miller, Ed L. (1992) *Questions that Matter: An Invitation to Philosophy* 3rd ed. New York McGraw Hill Inc.

Mishkin, Frederic (2001) *The Economics of Money and Banking*. Boston: Addison-Wesley World Student Press.

Moody, Ernest (1974) "Ockham, William of," *Dictionary of Scientific Biography*, Vol. 10, Charles Coulstan Gillispie (ed.) New York: Scribner and Sons.

Norton, David F. and Norton, Mary J. eds. (2000) *David Hume: A Treatise of Human Nature*, Oxford: Oxford University Press.

O'Connor J.J. and Robertson E.F. (1999) *Claudius Ptolemy, School of Mathematics and Statistics*, University of St Andrews, Scotland (http://www-history.mcs.st-and.ac.uk/Biographies/Ptolemy.html).

Outhwaite William. (1983) *Concept Formation in Social Science*. London: Routledge and Kegan Paul.

Peebles, James E., Schramm, David, N, Turner, Edwin L., Kron, Richard G. (1994) The Evolution of the Universe, *Scientific*

References

American October 1, (https://www.scientificamerican.com/article/the-evolution-of-the-universe/)

Popper, Karl (1968) *The Logic of Scientific Discovery*. New York: Harper Touchbooks.

Rabin, Sheila (2010) "Nicolaus Copernicus," *Stanford Encyclopedia of Philosophy*. http://plato.stanford.edu/entries/copernicus/.

Redman, Deborah (1993) *Economics and the Philosophy of Science*. New York: Oxford Press.

Reynolds, Jack (2016) "Jacques Derrida (1930-2004)" *Internet Encyclopedia of Philosophy*. La Trobe University (http://www.iep.utm.edu/derrida/)

Ricardo, David (1817) *Political Economy and Taxation*. [online] The Library of Economics and Liberty (cited March 11, 2001) Available from the World Wide Web: (http://www.econlib.org/library/Ricardo/ricP.html).

Ridley, B.K. (2001) *On Science*. London: Routledge.

Ritzer, George (1996) *Modern Sociological Theory*. New York: McGraw-Hill Company, Inc.

Robinson, Richard (2015) "History of Biology: Cell Theory and Cell Structure," *Biology Reference*, (http://www.biologyreference.com/Gr-Hi/History-of-Biology-Cell-Theory-and-Cell-Structure.html)

Roderick, Rick (1986) *Habermas and the Foundation of Critical Theory*. New York: St. Martin Press.

Rohlf, Michael (2010) "Immanuel Kant," *Stanford Encyclopedia of Philosophy* (http://plato.stanford.edu/entries/kant/)

Rosenberg, Alexander (1999) The Rise of Logical Positivism, in *Scientific Inquiry*, in Robert Klee, (ed.) New York: Oxford University Press.

Rouse, Joseph (1987) *Knowledge and Power: Toward a Political Philosophy of Science.* New York: Cornell University Press.

Rousseau, Jean Jacques (1981) "The Social Contract" in *Great Political Theories* Vol. 2, Michael Curtis Ed. New York: Avon Books.

Sandelin, Bo, Trautwein, Hans-Michael and Wundrak, Richard (2008*) A Short History of Economic Thought.* New York: Routledge.

Sears, Francis. Zemansky, Mark. and Young, Hugh (1979) *College Physics.* Massachusetts: Addison-Wesley Publishing Co.

Seeds, Michael A. (1988) *The Foundation of Astronomy.* Belmont, California: Wadsworth Publishing Co.

Skirry, Justin (2008) "Rene Descartes" *Internet Encyclopedia of Philosophy.* Nebraska-Wesleyan University (http://www.iep.utm.edu/descarte/)

Smith, Adam (1776) *An Inquiry into the Nature and Causes of the Wealth of Nations.* Chicago: Encyclopedia Britanica, Inc. Reprinted 1952.

Sparknotes (2015) *Plato: The Republic Book* VII 514a-521d (http://www.sparknotes.com/philosophy/republic/sections.rhtml).

Stewart, David and H. Gene Blocker (1996) *Fundamentals of Philosophy.* New York: Prentice Hall.

Strauss, Anslem and Corbin, Juliet (1999) "Grounded Theory Methodology: An Overview" in Bryman, Alan and Burgess, Robert (eds.) *Qualitative Research,* Vol. III. London: Sage Publication.

Suppe, Frederick (1999) "The Positivist Model of Scientific Theories," in *Scientific Inquiry: Readings in the Philosophy*